Walston's World

Walston's World

Oliver Walston

FARMING PRESS

First published 1999

1 3 5 7 9 10 8 6 4 2

ISBN 0 85236 527 6

A catalogue record for this book is available from the British Library

Published by Farming Press, Miller Freeman UK Ltd, 2 Wharfedale Road, Ipswich IP1 4LG

Distributed in North America by Diamond Farm Enterprises, Box 537, Bailey Settlement Road, Alexandria Bay, NY 13607, USA

CONTENTS

For Rose Nolen-Walston

INTRODUCTION

A strange thing happens to farmers when they pass the age of 40. They start to get restless. Some become active in the NFU, some get elected to the county council, some turn to drink and some get itchy feet.

I got itchy feet. Which is why many of the pieces included in this book describe my travels around the world. The opportunity arose when the editor of *Farmers Weekly*, a lovely man named Ted Fellows, suggested that I might like to drive round the EEC (as it was then called) and write a series of articles about the farming I saw. It must have been a success because he then invited me to do the same in Eastern Europe, which I did just as the Communist monolith was beginning to crumble.

Later, as my feet itched ever more itchily, I travelled to parts of the world where there was no farming at all. In particular, I fell in love with the Arctic.

But, however far I travelled, I always knew that my roots, my heart and my income were back on the farm at Thriplow. It was here that I began to see the agricultural world change more profoundly than in my previous 20 years of farming. The prosperity of the 1970s and 1980s ended with a bump. Gone were the days when my biggest worry was what colour new tractor to buy. Now it was how to survive. The number of farmers – and with it the power of the farming lobby – shrank each year, which meant that the voice of the taxpayer and consumer grew ever louder and more insistent. The CAP was reformed – or partially reformed – and instead of a very high supported price for my wheat, I received a vast (and very visible) cheque.

And then came the Mob, that bunch of self-anointed de-contaminators who, calling themselves Greenpeace and GenetiX Snowball, and supposedly acting in the name of democracy, smashed their way on to farms and destroyed crops. I myself am agnostic about GM foods. I don't know what effects they will have on the environment, or whether consumers will even want to buy them. But in order to answer these questions we must first be able to ask these questions. This means we must have proper trials. Of that I am certain.

I must thank the editors of *The Times*, the *Daily Telegraph*, the *Spectator* and *Farmers Weekly* for allowing me to print these pieces.

In the meanwhile at Thriplow we have just enjoyed the biggest harvest we have ever known. The yields were so vast that my father, who died in 1991, would simply not have believed them. It's what makes farming interesting, exciting and fun.

Oliver Walston
Thriplow Farm
August 1999

BARLEY BARONY

I HAVE SEEN THE FUTURE AND IT HURTS
1991

Package holidays for masochists are hard to find these days. I ruled out an in-depth study of slurry with David Richardson on North Humberside, an organic vegetarian hostel in the Brecon Beacons with Patrick Holden and a champagne-soaked balloon ride through Kenya – with Anthony Rosen providing the hot air. Eventually my shortlist was narrowed down to either a bed and breakfast in Baghdad or a camping holiday in Ethiopia. While I was pondering this difficult decision the telephone rang. It was Bill Smith, the producer of Anglia Television's *Farming Diary*. After 32 fun-filled years the programme was being axed and he had to spend his budget fast. Would I, he wondered, like to see how New Zealand farmers are managing without subsidies? Would I ever.

New Zealand is a bit like Ireland. Very green and full of friendly people speaking a sort of English with an impossible accent. It seems to have more sheep and fewer cows. The beer is pale yellow rather than dark brown. Another difference is that whilst Irish agriculture floats on a featherbed of subsidies provided by Sligo-born Ray McSharry★, New Zealand farmers have not seen a subsidy since they were abolished six years ago.

I had assumed, therefore, that New Zealand farmers would be both impoverished and unhappy. I was only half right. Big or small, sheep, milk or cattle, the men I met were, it is true, far from rich. In the past six years they had pared their farm businesses down to the barest essentials. Some had even gone beyond that and were no longer applying fertiliser, using herbicides or even making silage.

They admitted that they were leaner and hungrier than they had been in the bad old days of subsidies. 'In those days,' confessed a sheep farmer, 'we tried to produce as many lambs as possible. The name of the game was quantity. Today I am only concerned with making a profit and not just maximising my output.'

The statistics bore him out. Livestock numbers have fallen by 15 per cent since the subsidies disappeared. And yet at the same time the European Community has increased its production and is selling the surpluses on an oversupplied world market. 'How can we compete with 320 million taxpayers in Europe who subsidise your production of beef, lamb and butter and then once again subsidise your exports?' It was a frequently asked question to which I had no reply.

The hallmark of New Zealand farming today is simplicity. Dairymen are paid 5p per litre for milk and, unlike their British counterparts, do not have the comfort of a quota to use or sell as they see fit. No wonder they think hard before spending any money. The milking parlours were simplicity itself, usually 24 standing herringbones covered by a tin roof and unprotected from the elements. Into them trooped herds of Friesians which looked a bit unbalanced because their tails had been docked. This rather unpleasant practice makes life cleaner for the milkers but must make life hell for the animals when there are flies about.

Sheep producers are, if anything, in an even worse position. Over the past six years they have made just about all the economies they can think of, including the so-called Easicare system of lambing. In practice this means no care at all, so that the tough Romney-cross ewes are left to lamb by themselves on the hillsides.

With wool prices half last year's level and a fat lamb today worth £5, they are facing the worst year they have known for a generation. No wonder that New Zealand farm families follow the GATT negotiations with an interest which verges on the obsessive.

My second assumption, that the Kiwi farmers would be unhappy, turned out to be totally wrong. In spite of the sort of adversity which EEC farmers can only have nightmares about, I did not meet a single New Zealand farmer who wanted to return to the days of subsidies. From Owen Jennings, President of Federated Farmers (the NFU equivalent) downwards, they were adamant that they would never return to the days when 50 per cent of their income came from government handouts.

The reasons they gave were many and varied, but one in particular stuck in my mind. On a farmhouse terrace deep in the hills of the North Island, sheep farmer Cedric Percy paused over his teacup. 'When the government used to give money to farmers like me, the people in the towns began to dislike us. Now that the subsidies have stopped, we are once again respected by townspeople. And you can't put a value on that.'

Like early Christian martyrs in the catacombs of Rome, New Zealand farmers today seem to thrive on hardship and persecution. Their eyes light up and their voices become fervent as they explain how there really is life after subsidies.

And the fact that they are still farming proves it. The predicted wholesale bankruptcies never actually took place, land values did not plummet and suicides are still rare in the rural community. Of course life is tougher for New Zealand farm families than it used to be. People work harder, buy fewer new machines and drive older cars. But the crucial fact remains; New Zealand farmers can still produce butter, lamb and beef cheaper than anyone else in the world.

I have seen the future and it hurts.

> New Zealand still remains an island of unsubsidised agriculture in a world awash with subsidies. It would, however, be unrealistic to expect the EU to follow the New Zealand example and go cold turkey overnight. The 15 European countries, each with its own very different farming structure, would not and could not permit this to happen.

* Ray McSharry was EC Commissioner for Agriculture at the time.

HINDSIGHT MAKES ME HAPPY
1993

There hasn't been anything like it for 5,000 years. That was the day when a man (or probably a woman) first planted a seed in the aptly named Fertile Crescent. Since then, farmers have enjoyed prosperity and poverty as cycles have come and gone. But nothing in this whole period could possibly have equalled the riches enjoyed by British arable farmers in the 1970s and 1980s. Cereal growing was, as Lord Thomson once described the TV business, 'a licence to print money'.

Today, with the benefit of hindsight, I can look back nostalgically to an age when even the worst arable farmer could make money. The good ones made small fortunes and the excellent farmers became very rich indeed. Incompetence alone was not a reason to lose money. No, to do this with any certainty, you needed to be more than just incompetent. You also needed to have an extravagant wife and an addiction to roulette. Provided a man got up before midday, managed to scatter a few seeds in the autumn, and was prepared to harvest the crop the following summer, there was no way an arable farmer could fail to make a profit.

I entered farming in 1972. In the following decade I could not understand the old fuddy-duddies who were constantly harping back to the Bad Old Days of the 1930s. Farming, for me anyway, was a doddle. I should have realised that human beings' attitudes are invariably formed by their early experiences. Thus the men who had grown up in the 1930s remained cautious and pessimistic for the rest of their lives, while I – a child of the booming 1970s – was optimism personified. And for a decade or so I had figures to show that I was right.

My biggest problem was what to do with the profits I kept making. The solution was simple. I converted them into machinery by taking advantage of the 100 per cent tax write-offs we were allowed in those days. Of course, there was one disadvantage; the bank overdraft kept rising as I borrowed more and more to fund the purchases of bigger and bigger equipment. Tractors, combines, grain stores, grain dryers came in profusion. Just as I felt I should stop investing and concentrate on reducing my borrowings, along came Brussels with yet another scheme which gave me even bigger capital grants to encourage still more investment. As a result, I bought lorries and tractors with even greater abandon, safe in the knowledge that I was doing what the EEC wanted.

But I was not content simply to spend money on machinery. That was too easy. I also became greedy as I discovered the pleasures of owning more land. Over an eight-year period I managed to buy 300 neighbouring acres. As the price rose from £1,000 to £2,400 an acre, I knew I had made the right decisions. Not only did these purchases increase the size of my farm, but they also proved to be spectacularly good investments and land prices moved ever upwards. My shrewdness never ceased to amaze me.

Neither did my luck. There I was producing more and more wheat which nobody seemed to want. Yet this minor inconvenience did not stop Brussels from buying up the surplus, storing it at vast cost and then subsidising its export into a world market which was already awash with the stuff. It was the economics of the lunatic asylum as the floor price became the ceiling. The world of the Common Agricultural Policy was upside down.

Today, a decade after the glory days ended, I look out over a very different landscape. Instead of a prosperous countryside, I find myself surveying the wreckage of British farming as it struggles to come to terms with a harsh new world. The price of my wheat has remained static for almost ten years whilst the cost of everything else from wages to fuel has rocketed upwards. Instead of looking for more land to buy, I actually sold one third of the farm to reduce my debts. Instead of being urged 'to plant from fencerow to fencerow' I am urged to set land aside and plant nothing at all on 15 per cent of the farm. Instead of receiving subsidies to buy more tractors, I am today facing the prospect of making one third of the farm staff redundant.

I am not complaining. Far from it. For 20 years I enjoyed enormous prosperity thanks to the generosity of the government, and today it is my turn to suffer. Some of my fellow farmers are, however, less relaxed. They are blaming everyone but themselves for their misfortunes. It is, they claim, the government's fault for encouraging them to produce more food, the Common Market's fault for inventing the CAP or the banks' fault for lending them the money. This small minority of noisy and whingeing farmers deserves no sympathy whatsoever. Any farmer who is in trouble today because he cannot repay the money he borrowed in the good old days, has only himself to blame. He deserves no more sympathy than Mr. Robert Maxwell, whose greed also got the better of his judgement.

But the British public should pause before cheering the recent CAP reforms. It is true that the link between subsidies and production has at last been cut, so no longer will I inhabit the crazy world in which the more I produce the more subsidy I receive. Yet contrary to what Mr. Gummer claimed, food prices will not actually come down, the CAP will not cost less and the grain mountain will remain obstinately vast.

At the same time the landscape will change for the worse. Fifteen per cent of eastern England will grow weeds this autumn as we set aside our land in response to the new policy from Brussels. And as subsidies shrink and I am forced to compete with wheat producers from Canada and America, I shall have to farm like they do on the prairies, with few men and huge machines. So those naive and sentimental city-dwellers who imagined that the British countryside would somehow become more beautiful will be in for a shock. In the meanwhile farmers throughout Britain will be tightening their belts as they face a brave new world without subsidies.

> Here, for once, I was triumphantly right – or at least righter than John Selwyn Gummer. Food prices have not come down, the cost of the CAP actually rose and grain mountains are still a very real fact of life. Farmers like me hated setaside when it was introduced, but today we would be equally unhappy if it was abolished. That's the way farmers are.

A HAPPY FARMER
1995

I am a happy man. Happy not just because harvest has been quick and easy this year, happy not because prices are higher than last year, but happy because by Christmas I will receive a little brown envelope. The postmark will be from Guildford but the cheque inside for almost £200,000 will come from Brussels. It is my share of the Common Agricultural Policy.

It all sounded so sensible three years ago when the European Agriculture Commissioner, Ray McSharry finally succeeded in reforming the CAP. At that time grain, butter and beef mountains were growing ever higher and the cost of storing them – let alone exporting them at subsidised prices onto a glutted world market – had become totally insupportable. McSharry's idea was simple. Instead of Brussels guaranteeing a minimum price for my wheat, and buying any I could not sell profitably in the marketplace, this system would be gradually dismantled. Over a period of six years price supports would be progressively reduced as we farmers would inch nearer and nearer to the free-market world price. Of course, the farm lobby throughout Europe was horrified and began to mobilise the troops for a series of noisy demonstrations in Paris and Brussels. As usual, they predicted bankruptcy and desolation for Europe's farmers who would, they claimed, be crushed between falling prices and rising costs.

Politicians – at least those on the other side of the Channel – found the argument convincing and prevailed on McSharry to cushion the blow of falling prices. Which is why Brussels came up with the idea of area aids. This, in a nutshell, means that a farmer gets paid for every acre of crop he grows instead of for every tonne he produces. Brussels was triumphant. By cutting the link between the subsidy and production it had, to use the jargon, decoupled the payments. This was important not just to keep the taxpayers happy, but also to play our part in the GATT negotiations which were crawling towards a grudging agreement between Europe and the USA.

And so was born what British arable farmers call acreage payments. For every acre of cereals I grow, Brussels pays me £109. For an acre of oilseed rape I am paid £192 and for an acre of peas or beans £157. In order to buy an entry ticket to this new subsidy game, there was a price to pay. We had to set aside 15 per cent of our farms and promise to grow nothing on this land. But even here Brussels was in a generous mood. I now receive £138 for every acre I set aside, and on my farm this amounts to £40,000. Which is why travellers throughout the eastern side of Britain have recently noticed a lot of weeds growing where once there were cornfields.

The theory was good, but how did the scheme work in practice? The first effect was a tidal wave of hysteria from the farming community who complained vehemently that they were now being forced to fill up a long and complex document which took them many hours to complete. Few of them bothered to remember that in return for a few hours spent at their desks they would receive a cheque which would make most people weep with gratitude.

But farmers have never been known for their gratitude and they persisted in bewailing their fate.

Today, three years after Ray McSharry turned the CAP upside down, farmers have become accustomed to the new system and – as farmers always do – we have forgotten our original objections and instead have become devoted to the new regime. Meanwhile the world itself has changed. This may have come as a shock to the politicians, but it shouldn't have because if there is one thing certain about agriculture it is that it is a cyclical business. From the Old Testament's seven lean and seven fat years to the present day, we have seen these swings of fortune which make farming so predictably unpredictable.

Today it appears that we have left the cycle of surpluses which ran throughout the 1980s and 1990s and are now once again entering a cycle of shortages. Wheat stocks are at their lowest levels for almost 20 years and the world price has done the unthinkable – it has actually risen to European levels. It is not surprising, therefore, that setaside has been reduced from its original 15 per cent to 10 per cent, and some forecasters are predicting that it will disappear completely within the next two years.

All of which brings me back to why I am such a happy man this harvest. The crops have been smaller than last year so most people would expect me to be less than euphoric. They would be wrong. I just can't help thinking about that brown envelope which will plop through my letterbox at Christmas time. In it will be this year's acreage payment, and with it my continued prosperity will be assured at least for another 12 months. Unlike my wheat yields, the cheque will be about 25 per cent bigger than last year, which is good going when inflation is running at around three per cent.

The reason for this increase is twofold. Firstly it is structural; McSharry anticipated (quite reasonably as it seemed at the time) that the price of wheat would fall to world market levels and so my area aid cheque should rise to compensate me for this decreasing income. The trouble was that the experts in Brussels were no more accurate in their predictions than the people who speculate in the commodities exchanges throughout the world. The world price did not fall; instead it rose. But there is another reason why my cheque is so large, and that is because it is linked not to the pound sterling but the complicated green currency, which means that the weaker the pound, the stronger my bank balance.

But the brown envelope at Christmas will not be my only cheque this year. Sometime in the new year a smaller envelope will arrive. It will be for £6,000 and will be paid to me by the Countryside Commission as my reward for trying to look after the landscape of south Cambridgeshire.

Last year, troubled by the faint stirrings of what was once a conscience, I planted almost two miles of new hedges, repaired a lot of tattered ones, and sowed strips of wildflowers round the edge of some of our fields. I will repeat this process for the next four years and will, in return, be compensated in by the British taxpayer.

These brown envelopes represent what is wrong with agricultural subsidies today. In a sensible world the big cheque would be in return for looking after – and even improving – the environment. The smaller of the two cheques should be my reward for reducing the grain mountain and feeding a hungry world. But it will never happen. In the meanwhile I will look forward to Christmas and will try to remember to thank all those generous taxpayers who make my life so pleasant.

This was the high point (or low point if you happen to be a tax-payer and not a farmer) of the CAP lunacy. For two incredible fun-filled years the price of our products rose – due to world prices – while at the same time the size of our subsidy cheque also rose – due to Brussels' stupidity. My fellow farmers, as usual, hated me for pointing this out. Four years later, the EU Auditors, after much study and thought, produced a report which said that the whole system was flawed. What a surprise.

SELLING LAND
1990

I sold some land the other day. It was the most difficult decision I have ever made. Don't listen to anyone who says that it is the occupation rather than ownership of land which matters. They may be right in hard economics, but they fail to understand human nature. I had always assumed that the possession of land is the second most basic desire of the human race. The pleasure of owning the soil on which you grow crops, raise animals, walk and hunt is indescribable. Today, as I pass my 50th birthday, I must admit that the desire to procreate wanes while the need for more acres increases. A decision to sell land should, in theory, be a straightforward one based on a rational assessment of the facts and an intelligent guess about the future. In practice, nothing could be further from the case. To hell with budgets, cash flows and my fears about the CAP. What did I actually think about? I became obsessed with the pleasures of ownership. That was bad enough, but there were other unpleasant sensations which lingered in what I laughingly call my brain.

What would the neighbours think? Would everyone assume I was going bust? Would local merchants stop giving me credit? Did this mean I was a failure as a farmer? Or a failure as a member of the human race? All of these thoughts, and many others, joined forces to persuade me that I should sit tight and do nothing.

And yet the more I thought about the future, the less I liked what I saw. My farm, like so many others in Britain, had flourished mightily during the 1970s and early 1980s. In those days my biggest single problem had been knowing what to do with the profits which kept pouring in as both yields and prices moved steadily upwards. For a few years, I solved the dilemma by buying gigantic quantities of shiny new machinery. But it soon became clear that was not sufficient. Brussels kept throwing money at me, in the shape of the FHDS (Farm and Horticultural Development Scheme) grant and various other boondoggles. So I looked around for a quicker and more efficient way of getting rid of money. The solution was not hard to find. I bought land whenever it came up for sale on the boundary of the farm. I felt as if I had died and gone to heaven. Each year the farm got bigger. I have never met a farmer who would not like to have a few extra acres, and I was certainly no exception to that rule. All the clever people who advised me thought it was a wonderful idea. Parents, bank managers, accountants, united to assure me what I knew already – that the purchase of land was the best possible investment. And so it was for a few years.

The capital value of the land rose even faster than my profits, and within five years my investment had doubled. There were a few minor irritations, like the fact that my rent went up embarrassingly fast too, but I could live with that. So as the 1970s drew to a close I basked in the certain knowledge that not only was I a magnificent farmer, but I was also a stupendously shrewd investor. And then things began to turn sour. Prices went down, costs continued to rise and yields remained static. As if this were not bad enough, interest rates crept higher and higher. Nobody (not even a conceited farmer like me) could fail to notice that the whole political climate in the EC was changing profoundly. This was no minor hiccup which would all be over in a couple of years time. It was a profound seismic change which meant that once again British agriculture was entering a period of depression we had not known since the 1930s. I, being a product of the subsidised world of the 1970s, could not understand what was happening. Like a dinosaur facing the ice age, I was unable to cope.

One vision kept pushing itself to the front of my mind, and try as I would to dismiss it, I found it was starting to keep me awake at nights. Back in 1984 I had spent some time in Nebraska visiting bankrupt farmers; I was amazed to learn that bankrupt farmers were not necessarily bad farmers. Indeed the opposite was more often the case. The bankrupt farmers of Nebraska all had one thing in common, they had borrowed more money than they could afford to repay. Thus the realisation dawned on me that being a good farmer would not inevitably enable me to survive the 1990s. No. To survive I would have to be a low-cost farmer. And a low-cost farmer is a man who does not owe any money. That, of course, was not a comfortable realisation. Thanks to my land purchases, I was saddled with substantial debts. And so, over a period of five months I found myself spending more and more time talking to every expert I could find. Bank managers, accountants, land agents and consultants all sat patiently by as I cried on their shoulders. Their reaction was better than I had feared. I would, they reassured me, almost certainly survive anything the CAP might throw at us. Or if I did go bust, then so also would at least half the farmers of Great Britain. But that was where the good news ended. Survival, I was told gravely, might well be a very uncomfortable existence. Another year or two of drought, allied to lower prices and static costs, would mean that I should be under considerable pressure – and might even have to alter my lifestyle drastically. It would also mean that I would spend a lot more time worrying about things I was powerless to influence.

So after months of indecision during which I changed my mind every couple of days, I squeezed my guts into a knot and decided to sell one third of the farm. My objective was clear. I should end up owing no money whatsoever, and having a bit of cash in the bank with which to take advantage of any opportunities which might appear in the next few years. I would become a farmer who can survive in the real marketplace in the real world and not be dependent on subsidies. If wheat falls to £85 per tonne I shall still be in business. But I shall be a smaller farmer, with only two thirds of the acres I used to occupy last year. And that is not an easy fact to accept. In five years I shall look back on my decision and everything will be clear. People will not congratulate me for being shrewd. 'It was always obvious,' they will say in that smug manner which people use in hindsight. Or else they will laugh at that stupid fool who panicked. 'It was always obvious,' they will say. 'We told him so at the time but he didn't listen.' In the meantime, however, I shall sleep easy at nights.

> From a purely agricultural point of view it was a mistake to sell this land. I – and all of my advisers – were far too pessimistic about the future. You could even say we panicked. Yet in spite of this I'm glad I sold the land. It gave me a bit of liquidity which enabled me, for example, to buy a flat in London. So today I farm less land but have more fun. And that can't be a bad bargain.

DOWN WITH PLANT BREEDERS
1990

I don't have many heroes or heroines these days. No football players, no test pilots, no movie stars (Jamie Lee Curtis excepted) and definitely no politicians. But there is one man whom I admire. If you met him in the street (or, more likely, in a wheat field) you might not even notice him because he manages to blend in with the surroundings more than most people. His name is John Bingham and he is a (I almost said 'the') plant breeder. He is the chap who gave us Maris Huntsman and the whole series of wonderful PBI wheats which came out of Cambridge in the past 20 years. Every single arable farmer in Britain is a richer and, I assume, happier man because of John Bingham.

Why, I hear you ask, all this nauseating sycophancy? The reason is simple. I like plant breeders and am grateful to them. I also am terribly happy that plant breeders receive royalties on their varieties and can thus afford to pay John Bingham and his ilk. Please remember this fact if you manage to read to the end of this piece. A few weeks ago the British Society of Plant Breeders announced the new royalty rates for cereals. They did so without first talking to the UK Agricultural Supply Trade Association or the NFU, which was a trifle rude, but when you realise that most of the breeders are now owned by multinationals, maybe this is not surprising. They also decided that henceforth they would all charge different royalty rates instead of agreeing a single rate across the board. The reason for this is that they were terrified they might be accused of being a cartel. A not unreasonable concern when you remember that a group of supposedly competing companies had hitherto always agreed a single rate for their products. In addition to these two rather unimportant breaches with tradition, they also decided to increase the royalties on many (but not all) cereal varieties. It has been estimated that by doing so, the breeders gave themselves a pay rise of £1.25m a year. This, they explained hurriedly, was necessary because their costs were rising and only by paying themselves more could they possibly keep breeding new varieties for the future. Without this increase, they implied, plant breeding would grind to a halt, and no new varieties would appear. Advised by a very clever public relations outfit, they also managed to clamber aboard the 'green' bandwagon by claiming that they were now making great efforts to breed resistance into future varieties (as if they hadn't previously). The clear implication was that horrid polluting poisonous and unnatural chemicals would not be so necessary in the future. This is particularly ironic since most of the plant breeders are owned by multinational companies who make far more money flogging fungicides than ever they do selling cereal seeds. There is, I admit, some truth in the plant breeders' claims that they need every penny. Their costs are rising just like those of the rest of industry, and it is inevitable that they spend a lot of money today for a variety which may not appear for a decade. Or, worse still, will never even be released at all.

There are, however, two groups whom the plant breeders ignore when they plead poverty and hardship. Seed merchants are also finding life very difficult these days. Their costs are also rising and they too are worried about the future.

The recent spate of bankruptcies is proof of this. Reaction from merchants I spoke to ranged from 'appallingly short-sighted' to 'suicidal'. They were in no doubt that anything which increases the price of certified seed must reduce the amount of certified seed sold. The farmer too shares the same dilemma as the seedsman. He also would like an increase in revenue to help offset the lower price of grain and the ever-rising inputs. The breeders appear (outwardly at least) unworried. They reply airily that the new royalties will add only 2 per cent to the cost of growing cereals. Imagine what Ciba-Geigy Unilever, ICI and Shell would say if their suppliers arbitrarily increased prices by 2 per cent. There is, paradoxically, a group of men who will benefit even more than the breeders from these new royalties. The mobile seed cleaners will make a lot of money this autumn – and they will not pay a single penny in royalties. So, at a time when the price of cereals is falling steadily, when farmers and seed merchants are going bust, and when the entire industry faces a bleak future, the plant breeders have taken a long look at the golden-egg-laying goose and have decided to wring its neck. At this point a whiff of pure hypocrisy pollutes the atmosphere. Not content with increasing their royalties, the breeders lose no opportunity to abuse farmers who save their own seed. As well as being stupid, short-sighted and greedy, plant breeders appear to feel that they have a divine right to use the free market when it suits them (by raising prices) and to abuse farmers for operating the same system and choosing to produce their own seed. How anyone can contort himself into a position whereby it is sensible, desirable, prudent and right to charge more money for seed but at the same time it is 'immoral' for the farmer to save his own is something only a plant breeder (or maybe a Jesuit) can explain. In the meanwhile I, as a professional seed grower, shall find it hard to advise a fellow farmer not to save his own seed. Or should I, like the breeders, try to fill my own pockets and to hell with the consequences?

Those were the Halcyon (not a bad barley, by the way) days for plant breeders. Today – thanks in some measure to their greed – farm-saved seed takes a bigger share each year. Plant breeders insisted on raising the level of royalties at a time when their customers were being squeezed tighter and tighter. One day a plant breeder will learn about business in general and supply and demand in particular. By then, of course, pigs will be airborne.

THE GERMANS HAVE A WORD FOR IT
1994

The Germans have a word for it. They call it *schadenfreude*, which means taking pleasure at other people's misfortunes. Now it's true that most farmers suffer from this affliction a little bit, which is why we are slightly happy when we see our neighbour's field full of blackgrass or his new tractor broken down beside the road. But few farmers could have taken *schadenfreude* as far as me, which is why I have become an exceedingly unpleasant person.

The trouble is it's not easy being a genius, and even harder to remain modest and unassuming when I knew I was probably (no, dammit, certainly) the best grain seller in the world. As a grower of wheat I was fair to middling (I blame this on thin land and dry years), but as a seller of wheat I was up there with the Masters of the Universe on the Chicago Board of Trade. Wow! No wonder I had become bewitched, bedazzled and benumbed by my own brilliance when it came to selling the stuff that came off the combine.

A brief telephone call to some hapless grain merchant was all it took. My shrewdness, decisiveness and sheer talent must have left him breathless. Meanwhile my friends and neighbours (who are not necessarily the same people) would confess that half their harvest was still in the barn as the price fell ever lower. I tried to look sympathetic and muttered clichés like 'you can't get it right every time' while inwardly I chortled with delight.

And so it went on, with triumph succeeding triumph until something awful happened this spring. I started to believe my own propaganda. And that, as Mrs. Thatcher will confirm, is not a good idea.

It must have been sometime towards the middle of March when it happened. I had read the newspapers, talked to the politicians and (a big mistake here) listened to the grain trade. The CAP reform was well under way, with support prices falling and area payments rising. Intervention stores across Europe were bulging and my spies in Kansas, Argentina, Australia, Canada and the Ukraine assured me that a good harvest was on the cards. All this information was assessed by my computer-like brain and I decided to take a position. I would sell wheat forward for November. The first three phone calls resulted in offers of £89 per tonne but I wasn't born yesterday. I knew that was a lousy price and so stuck out for £90. Eventually one hapless merchant buckled under my pressure and bought at my price.

It was clearly a brilliant stroke and I lost no time telephoning my friends to let them know what the shrewd money was doing in Cambridgeshire that afternoon. They were obviously impressed because they were suitably and reverently silent on hearing the news.

And then the slurry hit the air conditioning. Using hindsight, it is now clear that the entire futures market had been waiting for me to make my move. Hundreds of little men in Chicago, Minneapolis and the City of London must have been poised for the news of what Walston had done. The price went up. Within a day or so of my making the big move, November wheat had firmed to £92. Within a month it had risen to around £98, and now, as the combines are about to go into the East Anglian prairies, the price is around £103 per tonne.

To say that I have egg on my face would not be truthful. I actually have egg on every bit of my anatomy, including those parts which egg does not normally reach. I have given up reading the newspapers or listening to the radio in case I learn what new and dizzy heights the price has reached. At nights my sleep is interrupted by recurring nightmares in which lines of grinning grain merchants shriek hysterically. During the day I sit slumped in the darkened farm office with a damp cloth over my brow fearful lest the telephone rings with more price increases.

And as if my marketing fiasco wasn't bad enough, the recent drought has made me less than optimistic about harvest itself. There will, however, be one agricultural crop in surplus at Thriplow this year. Is anybody in the market for sour grapes?

> This was probably the only article I have ever written which found favour with my fellow farmers. They were overjoyed to hear about my misfortunes. Since then selling grain has become even more fraught because the price is much more volatile than it then was. But for some inexplicable reason I haven't repeated this disaster – yet.

A SAINTLY BUNCH
1996

Farmers are, of course, a saintly bunch who go to church every Sunday, love dogs and children and always help with the washing up. But, let's face it, we do have just a few faintly unattractive characteristics. One of these is particularly noticeable at this time of year. If, having sold some grain, the price rises, the farmer is quick to blame the merchant and mutter darkly about how the multi-national shippers control the market. If, on the other hand, the price falls, the farmer is even quicker to take all the credit for his exceptional shrewdness and marketing skill.

Today I am a militant member of the latter group. I have long since forgotten how I cocked-up my wheat sales two years ago when I sold forward and watched the price rise by £20 per tonne. This harvest the welly is firmly on the other foot and I don't want anyone to forget it.

After the aforementioned fiasco I made a resolution that never again would I sell forward without taking out an option. This clever device may cost a few quid but at least it enables me to know that if the price falls I won't lose and if it rises I will benefit. Which is how last winter I sold a whole heap of wheat forward for November at £110 per tonne. Today, after a rollercoaster ride this year, the price is around £104 and I'm feeling smugger than a small bug in a big rug.

But what really makes me insufferably conceited is the fact that I sold 550 tonnes of feed wheat during the first week of August for an average of £118 per tonne. It's now safely delivered and – if the small but perfectly-formed Richard Whitlock of Banks Agriculture is to be believed – the cheque is in the post.

So – on the financial level at least – harvest started well. The snag was that our land needs a lot of rain and I wasn't at all optimistic about the yields. The first two fields, both second wheats on thin land, bore me out by producing a scrawny 50cwt per acre. But then something strange began to happen. The yields climbed higher and higher until a 75-acre field of Rialto managed to give us 319.9 tonnes over a weighbridge at 15 per cent. For those who find arithmetic difficult, that's 85cwt/acre. Which is not bad for a milling wheat on land that can't remotely be called heavy. It must, I told myself modestly, be a reflection of my incredible skill.

And the harvest itself has been fun and exciting. A few years ago we operated four combines and this year we are down to just one. But what a machine. The Claas Lexion 480 is to combines what a Boeing 747 is to planes – and travels at about the same speed too. We used it last year and were, I admit, disappointed with its output in dry conditions, but this year it returned to us with a bigger engine, various modifications inside, a GPS device to produce yield maps and – glory of glories – a 30 foot header. It may be a nuisance to move from field to field but it usually thrashes 35 tonnes per hour and sometimes a lot more. No longer do I have to look longingly at the American combines which cut their way north from Texas to Montana. This year we've got one ourselves and it's as exciting as my first Dinky toy was 50 years ago.

So what with good prices, good yields and a good combine, I'm a very happy little chap. And as the great British public knows well, a happy farmer is a rare farmer.

> On reflection, this is a pretty smug piece. However, I suppose it redresses the balance and shows that once in a while I do get things right.

PRECISION FARMING
AUGUST 1998

There has only been one real revolution in my farming lifetime; the intensifica-
tion of cereal production which took place in the late 1970s. They were exciting
times as we came to grips with new varieties like Maris Huntsman, new
practices like tramlines, and even new problems like how to export the surplus
wheat which we started to produce. Today, nearly a generation later, it is easy to
forget how much courage it needed to adopt all the new techniques which
today we take for granted. We didn't only have to invest in new machinery,
we also had to rearrange our brains and forget much of what we'd learned about
growing wheat so we could start to treat the crop almost as a market gardener
treated his lettuce plants under glass. Today another revolution is just beginning
to appear, and its effects will I guarantee – be more profound than anything
British agriculture has seen since Jethro Tull and Turnip Townshend. I refer, of
course, to precision farming. The snag is that precision farming has already been
so over-hyped that most farmers are thoroughly bored by the subject, and I can't
say I blame them. Yet the fact remains that when an unmanned piece of farm
machinery can locate itself on a field to within a few centimetres, the possibilities are limitless. Among my less pleasant characteristics is the fact that I suffer
from chronic impatience. Everything should have been done yesterday – or
preferably last week. Which is why I find the precision farming revolution both
fascinating and frustrating. For the past three years I have been visiting the
agricultural shows where eager young men spouted the virtues of GPS (Global
Positioning Systems) and showed me maps and charts. I examined quad bikes
with weird aerials, combines with strange gadgets, and fertiliser spreaders with
exotic control systems. Each exhibit was accompanied by glossy brochures and
diagrams proving conclusively, that with GPS-controlled equipment I could cut
my costs and increase my yields.

As a certified gadget-freak, I became nearly as excited as the exhibitors themselves. Until, that is, I returned to the farm and realised that it was all so much
froth. For the first two years I had been looking at what the computer fraternity
calls vapourware. But last year things began to change. The leaders of the GPS
revolution did at long last manage to produce things which worked. Sort of.

For the past two harvests our combine has been able to produce beautifully
coloured yield maps which impress my friends but are of absolutely no practical
use whatsoever. Likewise the office computer can produce field maps which are
so professional they look as if they have been printed by Ordnance Survey
themselves. So what? All these gadgets, it is true, now at last do what they were
designed to do, but they still lack an absolutely crucial characteristic – they can't
actually speak to each other. If you were setting up a farm for the first time this
might not be a problem, but since none of us can afford to go and buy all new
tackle, we have to make do with what is in the barn today.

My frustration became so acute I could stand it no longer. So I summoned a
meeting at Thriplow to which I invited some of the people who manufacture
the machinery we use today. I sat them down in the farm office and outlined my

plan. This harvest, using our existing Claas Lexion combine, we shall produce yield maps for three different fields. After harvest the people from Soyl will appear with a GPS-controlled quad bike and make us maps of the same fields. The resulting PCMCIA cards will be slipped into the farm computer where our Farmade software will produce a single map on which we will see both sets of data superimposed.

Based on this information I shall then make some profoundly intelligent management decisions about fertiliser applications this autumn. This data will again be put on to a tiny PCMCIA card and pushed into a slot in the LH control system which by then will have been fitted to our existing Bredal fertiliser spreader. As a result P and K application rates will be varied automatically and accurately. Racal Landstar, whose equipment links these machines to the satellites, will provide a portable system which will enable us to walk to any point on these fields which has been pinpointed for further examination.

It all sounds so simple. But course it won't be. There will inevitably be snags, frustrations and good old-fashioned cock-ups. But if we could achieve this objective using the machinery on the farm today, we will have taken the first faltering steps towards the revolution that (trust me) is hovering over the horizon. Which is why I'm more excited today than I have been for a long time.

November 1998

There will be, I wrote in the first article, 'snags, frustrations and cock-ups'. And that just about sums up my first harvest as a gung-ho, red hot and extremely naive precision farmer.

If truth be told I have only myself to blame. Gadget freaks, like alcoholics, find it very difficult to say no. There I was one sunny afternoon being shown round the Claas factory by a very, very, very, senior executive. He had summoned the bosses of the precision farming department who were keen to show me their latest inventions. Among these was a device which they called AgroCom. It consisted of a compact and detachable computer built into a colour monitor. But what really excited me about AgroCom was not simply its good looks, but the fact that it could be moved from combine to tractor or quad bike. Was it I asked (trying to conceal my excitement) ready for on-farm use? The reply was a straightforward 'Ja'.

On my return home I phoned the Claas importers and told them about my new discovery. To my surprise, they did not share my enthusiasm. On the contrary, they pointed out the old Cebis system I had used last year had been perfectly reliable and thus they were distinctly uneasy about providing me with the latest technology until they were completely satisfied that it worked. I persisted and persisted and persisted. Three weeks later a shiny new AgroCom was fitted to our Claas Lexion 480. The combine cab now began to take on the appearance of Regent Street at Christmas time. Coloured letters and symbols on the AgroCom made the old Cebis seem very dreary. I was really looking forward to harvest. And then it all started to go wrong. After we had completed the first field of winter barley I seized the PCMCIA card, rushed back to the farm office, shoved it into the card reader and waited impatiently. A minute later I found myself looking at half a field. The other half had vanished into inner space. After that things went from bad to worse, and nobody at Claas seemed to know why. Experts arrived from Germany, spent the day fiddling and went home full of smiles, assuring me that the problem had been solved. It had not. Indeed the yield maps became even worse, with long straggling lines which looked as if a demented chimpanzee had been playing with a geometry set. By now harvest was drawing to a close and I was becoming desperate. Finally, and with much reluctance, I agreed to have the old Cebis system fitted again, and we eventually managed to produce yield maps for the last three fields we combined.

Meanwhile the other aspects of the great precision farming experiment were also taking shape. Peter Henley from Farmade installed his latest prototype mapping system on the farm computer, but worried me slightly by saying that the software was far from being bug-free and we might still eventually have to use the current system. The people from LH Agro came to fit their GPS-controlled system to our Bredal spreader but, having removed the old-fashioned land wheel drive, they went away leaving us with an unusable – but very high-tech – machine. As the mists of September started to make the stubbles soggy, the man from Soyl arrived with his quad bike and whizzed all over the three fields which we had selected. Two weeks later his boss Simon Parrington presented me with a folder in which were stunningly beautiful maps showing the P, K, Mg and pH levels in glorious Technicolor. I was fascinated to see how great the variations really were but was equally appalled to realise that in the past we had simply dumped the same amount of fertiliser across the entire field.

Today, as we approach the climax of the operation I am feeling a good deal less optimistic than I was when I dreamed up the project. The yield maps were a fiasco this harvest, though this was largely my own fault. The spreader is still in pieces in the barn. The computer mapping software may or may not be quite up to the job of producing instructions for the spreader. What had seemed so straightforward back in the spring looks distinctly dodgy today. But then, as I try to reassure myself, it probably took Harry Ferguson a year or two to get his linkage straightened out, and even the Wright Brothers didn't get airborne at the first attempt. Meanwhile there's still one more problem to cope with: an information overload. I now know precisely how and where the three fields yielded, and I also know where the P and K are high and low. But what on earth should I do with all this knowledge? The next article will reveal all.

APRIL 1999

The last of these articles ended with me quivering with excitement as I waited for Soyl's quad bike to appear on the farm and start taking soil samples. The appointed day dawned fine and the quad bike duly appeared on a trailer behind a Vauxhall Vectra. I watched with ill-concealed excitement as the operator set up his satellite navigation antenna, booted up his laptop computer and disappeared off across the stubble to start taking the first of his umpteen samples.

It only took him a morning to cover 80 hectares, and all he had to show for it was a plastic bag full of soil samples and a laptop full of data. After this exciting moment weeks passed during which I heard nothing. Eventually I received through the post a computer disk with instructions on how to transfer this data onto our Farmade mapping programme. Now I yield to nobody in my enthusiasm for computers, but I was completely flummoxed by the instructions. This was without any doubt the least friendly software I had ever used and after a morning's frustration I eventually gave up. No problem, said Simon Parrington of Soyl, he would transfer all the data onto a PCMCIA card and our LH Agro hardware on the Bredal spreader would be able to do the rest. Or so he said. It all sounded so convincing but by now I was becoming cynical.

A week later Simon Parrington turned up in the farm office, armed with a fistful of beautiful maps displaying the phosphate and potash levels in glorious technicolor. The problem was that they showed absolutely no similarity to the yield maps which the combine had produced last harvest. But there was no time for such details because we were already way behind schedule for base fertiliser spreading. Soyl had not just provided us with maps; they had also worked out how much P and K we would need for next year's cropping. Thus for the first time in nearly 30 years I found myself ordering straights instead of blends from my friendly local fertiliser supplier. In the meanwhile LH Agro turned up to fit the necessary hardware to the tractor, including a satellite receiver and the computer control box for the spreader.

And that is where the problems began. For the past ten years we had used our Bredal spreader very effectively. Armed with a weighbridge and an instinctive feel for machinery, Ted King invariably managed to get our application rates to within a few kilos per hectare of the target figure. This year he returned from his first load looking miserable. It appeared that the new hardware, which dispensed with the old land wheel drive, meant that the traditional settings were wrong. Very wrong indeed.

A week later, after many experiments, there was a smile on Ted's face. He had worked out the new settings and was now less than one kilo per hectare off target, which is as good as one could possibly expect. What was even better was that the GPS system was working perfectly so that in some parts of the field the spreader stopped completely whilst at other locations it was putting on very great quantities. The realisation that the machine was not being controlled by Ted but by a computer and a satellite gave us an eerie feeling. It was almost as if we had peered into the future and had caught a fleeting glimpse of how farming would be done by the next generation.

So was it all worthwhile? After a few nervous moments in the autumn, there could be no doubt that the experiment had been a total success. Using our existing machines we had succeeded in applying variable rates of fertiliser based on extremely detailed soil samples. The hardware is available and can be bolted on to existing tackle and the actual control mechanism can be operated by a skilled tractor driver. The mapping software, however, is more of a problem, and should really only be used by experts who know exactly what they are doing. The day will, however, eventually come when even a dumb farmer like me can work it, but today this would be unwise.

The combine yield-mapping information is, I am forced to admit, a bit of a luxury. The maps may be pretty but it is hard to see what to do with the information they contain. The soil analysis maps are, however, absolutely wonderful and make precision farming totally worthwhile. Two generations ago my father's farmworkers dumped heaps of manure on a field exactly one chain apart; these were then spread by hand. Years later I discovered the joys of soil analysis and applied the fertiliser at different rates on different fields. Today we have moved on another stage and, thanks to GPS, we can now apply fertiliser precisely where it is needed across every field. This is good news for me because I save money, and it is good news for the environment because we no longer spread the stuff all over the countryside.

Precision farming may have started slowly but it is here to stay. I can't wait for next autumn when we start spreading again.

THE AGONY OF IACS
1995

Among my many unattractive habits (which include gluttony, foul language and laughing at my own jokes) is the fact that I make up slogans and then proceed to believe them. An example of these is 'we'll never go broke if it rains'. I live in the driest part of England and farm on some pretty light land so this particular slogan was one I felt particularly proud of. It seemed to sum up in the pithiest possible way, how dependent we are on rain. Today, after the wettest winter I can ever remember, I'm giving up slogans for Lent – and maybe longer.

Another reason I'm not in the best of moods is that the IACS forms appeared on my desk last week. Now unlike most farmers, I am not an IACS whinger. On the contrary; it is the easiest money I have ever earned (earned?) in my entire life. It takes me about three hours to fill in the forms, and in return I received a cheque last year for £150,000. That is, after all, almost as much per hour as the head of a privatised industry gets paid. No wonder I feel deeply nauseated when I hear of yet another farmer complaining about the time it takes for him to fill in these forms and how unnecessary the bureaucracy is. Any idiot who still feels this should remember that the whole system is voluntary. If he really dislikes it so much he should just tear the forms up and throw them into the wastepaper basket. But to collect your acreage payments and at the same time complain about the system makes the public even more certain that some farmers are ungrateful parasites.

The reason I am uneasy about the IACS forms is that I almost came unstuck in a big way last year. I filled in the forms, took them to my local MAFF offices, had them stamped and went home feeling about as smug as smug can be. A few months later I got a phone call from MAFF; they proposed to do a spot check on the farm and thought I should know in case I saw people wandering all over the place pushing measuring wheels. Be my guest, I replied breezily.

Harvest came and went and the first of the oilseed rape cheques arrived. I didn't have a care in the world until one day the phone rang again. It was a man from MAFF who wondered if he could come and see me. There were, he said, 'a few problems' he'd like to discuss. There was something about the tone of his voice – just the slightest hint of menace beneath a polite veneer – which made me feel a tiny bit uneasy. You get the same feeling when a Customs Officer stops you in the Green Channel and asks to look in your suitcase.

Most of the problems were trivial. They disagreed with me about the size of some fields but the differences were minute; a fraction of a hectare either way. I was beginning to think the whole interview was a formality when the man from the Ministry cleared his throat. 'Unfortunately,' he began, 'we have come up with what looks like a serious discrepancy.' And serious it certainly was. I had muddled the measurement of two adjoining fields so that one was 2.5 hectares too big and the other 2.5 hectares too small. As a result I had underclaimed for my wheat area but had overclaimed for my setaside. I gulped and started to apologise. I could see that my excuses were having no effect. It was time to cringe and grovel. My humiliation left the civil servant utterly unmoved. The penalties for incorrect declarations, he pointed out, are absolutely clear. Indeed, if you over declare your setaside area by more than 20 per cent you will lose all your IACS payments. Everything. The whole shooting match.

As it happens – and by sheer good luck – my eventual setaside area still turned out to be fractionally over the necessary 15 per cent of the whole farm and so I was not fined. Instead, my setaside payments were reduced by £2,335 and I emerged with a sharp rap on the knuckles. This year I will treat the IACS forms with even greater care and will make sure that, just to be on the safe side, I set aside a few acres more than absolutely necessary. In the past two years I've got quite attached to that little old IACS cheque. In fact I've become addicted to it.

> I may have become addicted to the IACS cheque but I still hate filling out the forms. As someone who failed O level maths, I find all that addition and subtraction very difficult. But I shouldn't complain because the prospect of that cheque still makes me go weak at the knees.

STUPID MAN
1995

Like most stupid men, I have very strong opinions. What is even worse, I like the sound of my own voice, particularly when it is pontificating on an important topic. Thus, if you had asked me about farm labour six years ago, I would have had no doubts whatsoever. At that time we employed 16 people on 3,000 acres, which included a suckler herd of 85 cows and a flock of 475 ewes. My first reaction would have been to state categorically that one man per 200 acres was a magnificent achievement which showed how efficient we were. If you had persisted, and had managed to survive my barrage of conceited certainty, I might just have softened my position. I might have been forced to admit that the farm could, in theory at least, be run with one fewer man.

If, being the intelligent and insightful person I know you to be, you had suggested that, faced with our high fixed costs and the dismal outlook for farming, I should cut our wages bill in half, my reaction would not have been a pretty sight. I would have become unpleasant and inflexible; showing all the signs of being dogmatic in the manger. I would have started by announcing that to run the farm with eight people was simply impossible. Convinced by my own infallibility, I would have quoted figures from John Nix and Michael Murphy (the two great gurus of agricultural economics) to show we were doing well. I would have gone on to claim that – if ever the labour force really were halved – the farm would go to pieces, the buildings would fall down, the contractors' bills would be huge, the overtime would become intolerable and – most arduous of all – my job would become impossible as we fell further and further behind with the jobs. Utterly convinced my by own brilliance, I would have collapsed into a steaming heap of self-righteousness with a glass of bourbon whiskey in my sweaty hand.

Today, six years later, I realise how totally wrong I was. We now employ not eight, but six people (albeit without livestock). The crops and the buildings look no worse than they ever did, we use a contractor only to spread P and K – exactly as we did six years ago. The overtime bill has not gone up much, but since everyone is on salary (except Tony, the Youth Opportunities Programme student), there is no overtime paid in the first place. Most surprising of all, my job – far from becoming more difficult – is actually easier. It is far simpler to manage six people than sixteen.

You may detect the faintest signs of smugness here. And you are probably right. There's no doubt I'm proud of (and a touch smug about) the six men on the farm today. Dave, the manager, Ted the foreman/fitter, Stan, Brian, Lindsay and Dick are about as good as any farmer anywhere could possibly expect. With an average age of 33, they are all tough, bright, keen, energetic, sensible and – above all – flexible. They are, in short, not locked into the old habits and old attitudes which so cripple both farmers and farmworkers alike.

This means that we can do something I never thought possible six years ago. Once a year we all meet in the farm office to go over the annual accounts. Nothing is hidden. They all know what profit – or loss – was made, how much money was spent on wages, and how much money I pay myself.

There are other advantages of having a small labour force which I had never anticipated. Early in January we all jumped into two cars and went over to the Claas factory in Germany to watch our new combines being built. Two nights on the ferry and one in Westfalia, not to mention an evening in Amsterdam, were great fun. They didn't do morale any harm either. This expedition would not have been practical if we had needed to transport 16 bodies across Europe.

I tell you all this not, as you probably assume, because I'm pleased with myself, but because an awful realisation dawned on me recently. Were you to ask me today whether it was possible to cut down on chemicals and fertilisers, my reaction would be the same as it was six years ago when talking about labour. I would splutter and chunter about how careful we are today. I would reassure you that the days of prophylactic spraying, when we chucked everything on before we saw a sign of disease, are long since gone. I would cite trials results which show that everything we do is cost-effective. My voice would grow louder and my face redder, as dogmatism replaced doubt.

But however similar my outward reaction might be to that of six years ago, somewhere in what I laughingly call my brain a worry would be germinating. It would force me to remember how wrong I had been about labour. It would also force me to wonder if today I am probably (not certainly, you understand) extravagant with fertiliser and agrochemicals.

So come back and visit me in 1996. I shall be six years older but not a lot wiser. I will have somehow forgotten that back in 1990 I was so vehement that we were using sensible amounts of fertilisers and fungicides. Instead, I shall be singing a very different song about how extravagant we all were back in the bad old days.

I won't have turned organic – unless I've won the football pools in the meanwhile – but I'm fairly certain I shall be using far fewer inputs than I do today. And yet the yields, crops and countryside will look the same. Only the gross margins will be different. Of course, until 1996 arrives I'll deny I could possibly use fewer chemicals. This should not be a difficult task; I've always found myself very convincing.

> Nearly ten years later I find that we employ four men to farm my 2,000 acres. We do – as I predicted – use less chemicals and fertilisers, and I haven't turned organic (because I haven't won the football pools). Today I face a future which even ten years ago I could not have foreseen. Genetically modified crops and satellite-controlled machines are just beginning to appear on the horizon. It's all a bit scary.

FARMING IN BRITAIN

COCK-UP BEYOND ALL BELIEF
1989

I had been expecting it for over a year but, like all bad news, it was still a nasty shock. On 31 July the new Minister of Agriculture, John Selwyn Gummer, made the announcement. His Ministry was, at long last, getting to grips with the problem of nitrates in the groundwater. Twelve parts of England had been selected as 'Nitrate Sensitive Areas', and a further eight were described as 'Candidate Areas for Intensive Advisory Campaign Only'. The word 'only' seemed particularly sinister. But worse was to come. One of these eight areas consisted of 2,000 hectares at Fowlmere, the neighbouring village. I braced myself for the worst. The phone began to ring as friends, acquaintances and enemies extended their deepest condolences (and tried not to show their glee).

The next day I was waiting for the postman. Sure enough there was a brown envelope with a MAFF postmark. The epistle began 'Since it appears that you may farm land in one of the areas which have been selected for action, you may care to have some further details of the arrangements.' That was an understatement.

The letter described how I would receive a free visit from an ADAS adviser next spring or summer as part of an advisory campaign. This adviser will 'discuss farming practices which will help to reduce the risk of nitrate leaching, such as the rate and timing of fertiliser applications – but no compensation will be paid for following that advice.' It went on to say that if I wanted 'a little more detail or discuss the matter with a member of the Ministry's staff' I should ring Kings Lynn 763486 and speak to a Mr. A (to protect the innocent I shall not reveal his name).

I may have waited half a millisecond before picking up my phone and asking for Mr. A. I found myself speaking to a civil but clearly uncomfortable civil servant. What, I wondered, was the ADAS man going to advise me? Was he going to suggest that I used less nitrogen, or would it simply be a matter of timing? 'I'm afraid I can't help you on that one,' said Mr. A. 'We haven't been told what form the campaign is going to take yet.' He went on to explain that the areas were all pilot schemes designed to provide the information which could later be used on far larger areas in the rest of the country. This, I confess, gave me a slight twinge of satisfaction. I was slightly less unhappy about being used as a guinea pig if I knew that my fellow farmers were also due to endure the same punishment at a later date.

But back to the telephone conversation. What would happen, I asked Mr. A, if I refused to take the ADAS man's advice? Here too he was unclear. 'I can't really say. On the one hand it is a voluntary scheme and you are at liberty to do what you like, but on the other hand the new water legislation does contain compulsory powers which could always be used if the need arose.' This did not seem particularly helpful. 'I know,' replied Mr. A. 'But please understand that we in the

Ministry of Agriculture are here not to make the rules but simply to administer them and nobody has yet told us what we should do. It is, in a way, a pity that the announcement was made without first giving us our instructions isn't it?' I agreed.

Undeterred by Mr. A's impregnable ignorance, I continued on my mission of discovery. How would they monitor the results in my area? Would they, for example, sample the water after a year and – based on the new nitrate level – issue even more draconian orders? 'Here again we are not certain,' conceded Mr. A. 'I do appreciate that the water from any particular borehole need not necessarily have come from an adjacent piece of land. It is all very complicated isn't it?' It certainly was.

The conversation ended with protestations of undying affection and gratitude – and both of us agreeing to keep in touch.

The announcement of these Nitrate Zones has all the hallmarks of another CUBAB★ perpetrated not this time by Ministry of Agriculture civil servants but by the politicians. It is, in other words, a crude political gimmick which has been announced prematurely simply to show the British public that 'something is being done'.

And it may well work too. It may, I suppose, fool a few voters that at last the Conservative government is acting rather than talking about the whole problem of nitrates in drinking water. But it hasn't fooled men like Mr. A who are supposed to administer the scheme and yet have not the faintest idea of how it will work.

If only John Selwyn Gummer had restrained his short-sighted, short-term political tendencies and made sure that the details had been worked out before he made the great announcement, all would be different. But that is not the way politicians work.

I feel sorry for Mr. A and his colleagues in the Ministry. But not half as sorry as I feel for me.

> Nitrate Sensitive Areas are still with us, but for some reason they don't seem important any more. In fact, until I came to write this update, I had completely forgotten that part of my farm is in one of these areas. Maybe they were simply a political gimmick dreamed up by the fertile mind of Mr Gummer.

★Cock-Up Beyond All Belief

DEFENCE AGAINST FARMER BASHING
1989

Fifteen years ago a new sport was invented; it was called farmer bashing. Since then I have watched the fun from the front row of the stalls. Television documentaries, articles and even cartoons played their part, and there were even moments when, in a fit of masochism, I gleefully joined in the game myself. Unlike many of my colleagues, I have never believed that I should refrain from criticising farmers. Quite the contrary. In those early days we farmers were the softest of soft targets. Not only did we deserve a lot of the criticism, but we did nothing to defend ourselves. We were too busy making lots of money. It was, as the Americans say, a bit like shooting fish in a barrel. But today something deeply unpleasant and very different is happening. No longer are we seen as mildly selfish people who pull down hedges, put up 'Keep Out' signs and complain a lot. Today we are being depicted as polluters of the countryside and poisoners of the population. Aided and abetted by the Ministry of Agriculture, we farmers will do anything in our ceaseless and utterly single-minded pursuit of profit. Or so the story goes. It all came to a head, of course, with the salmonella scare.

There was, it is worth remembering, a brief moment – probably no more than a couple of days – when farmers in general, and Simon Gourlay in particular, were national heroes. We had stood up against the slanders of Edwina and had seen her off. The TV news programmes were full of decent and hardworking poultrymen who were on the verge of bankruptcy. And yet within minutes of Mrs. Currie's★ resignation, the world returned to normality. Once again farmers were the villains of the piece. Not only had egg producers been poisoning the population, but the NFU was back to its old tricks as that shadowy, all-powerful and unscrupulous lobby which manipulates politicians and civil servants like marionettes. Since then the attack on agriculture has become louder and ever more hysterical. Scarcely an evening goes by without Richard Body★, Professor Lacey★ or a man from the London Food Commission appearing on the telly to inform the world that farmers are wicked and greedy and the Ministry of Agriculture is incompetent and weak. As a result of the above, the food we eat is at best dangerous and at worst deadly.

Throughout this barrage a small group have sat back and enjoyed the spectacle hugely. The organic farming lobby has lost no time in assuring the public that none of this would (or could) have happened if only the farmers of Britain had not been seduced by the dreaded agrochemical multinationals. Free-range, organically fed chickens, it implied, are magically not infected with salmonella. Listeria, it seems, does not affect organic produce. A few minor inconveniences, like price and availability, are conveniently overlooked. And the overwhelmingly important fact that organic farming is a rich man's hobby is never mentioned at all. As the hysteria mounts and the misinformation (I almost wrote 'lies') increases, there is a deafening silence from the entire agricultural industry. The NFU, God bless it, has for once done a bit (albeit not very effectively). But what have we seen from the people whose lives depend on farming quite as much as our own? The agrochemical industry has maintained an embarrassed silence, UK Agricultural Supply Trade

Association has hoped the problem would go away, the machinery dealers have clearly felt it has nothing to do with them, and so on and so on. These poor, pathetic, short-sighted and, above all, cowardly idiots have apparently not realised that unless someone stands up for the industry we may all go down together. In the meantime UK farmers, with incomes down 25 per cent on last year, batten down the hatches and numbly wait for the next assault. And the next assault will come on 7 March when the Duke of Edinburgh, as Chairman of the Council of Food and Farming Year, delivers the Dimbleby lecture on 'Living off the land'.

It will certainly be an imaginative, thoughtful and possibly even provocative speech. But I have little doubt that there will be at least a few sentences which, if taken out of context, will serve the purposes of the anti-farming lobby. The national press will, of course, publish these under large and hysterical headlines. And thus the campaign will continue. So what can be done? Plenty.

● Join the NFU if you are not already a member. I know I've spent a lot of time attacking the NFU, and it is still less than perfect. But it's our only chance and we'll need it over the next few years a lot more than we've needed it since World War Two. Any farmer today who really feels he will do better outside the NFU is a fool. And, worse still, a selfish fool.

● Get out into the real world and talk to non-farmers. And I mean talk, not just whinge. There's no point in telling them about how badly off we are now and how our incomes have fallen. They won't be sympathetic, and I can't really blame them either. Tell them a few facts of life. Tell them, for a start, why you farm as you do.

● Don't be afraid to write to your local newspaper or radio station. But again, don't whinge.

● Get involved in local politics. Or even national politics. But don't whinge.

● Tell the man who sells you machinery or chemicals or concentrates that his livelihood is on the line too. He can no longer just hide behind the rest of us and hope the problem will go away. His bosses should be doing all of the above.

● Above all (in case you still have not got the message) resist the temptation to whinge. We farmers are often our own worst enemies by constantly demanding public sympathy from an unsympathetic public. Our job is to explain, not complain.

> I feel about farmer bashing the same as I do about people being rude about my family. It's fine for me to be rude about my parents but I'm damned if I'm going to let any old outsider be rude. Which is why I don't mind bashing farmers myself but I get very irritated if someone else does.

★ Edwina Currie was a junior minister who caused a tidal wave of anger among farmers by suggesting that the risk of salmonella poisoning made British eggs unsafe to eat. Richard Body, an eccentric Tory MP and Professor Lacey, an academic, all pitched in to whip up public hysteria. For once I was on the side of the NFU.

SUGAR BEET SECRETS
1997

'I can't give you my name,' the telephone caller began, 'because it would cost me my job.' There was a pause and I heard papers being rustled. I assured the caller that I had no desire to know his name and I would not even attempt to find out. We talked for about 15 minutes and at the end of it he promised to send me some figures which 'should make all sugar beet growers very angry indeed.'

A week later a brown envelope arrived on my desk. The address was handwritten in block capitals and the postmark was unreadable. Inside I found a single sheet of paper which had been neatly typed. It contained all of the figures which we had discussed on the phone.

My anonymous friend's document is – just as he had predicted – rather interesting. His fundamental point is that British Sugar is 'ripping off' growers to the tune of £23 million a year. Since there are 9,700 sugar beet growers in Britain today, this is costing the average grower £2,300 each year. On the face of it this is an incredible claim but my anonymous caller produced the following facts to substantiate his claim.

Eight per cent of all beet delivered is never actually paid for. This is because British Sugar claims that this proportion is the crown and – under the terms of the Interprofessional Agreement – must be deducted from the total. But in fact these crowns, which amount to 712,000 tonnes and have a sugar content of at least 13 per cent, are all used to produce sugar. From them British Sugar manages to extract 80,882 tonnes of sugar.

The beet growers of Britain are thus, claims my friend, giving to British Sugar free, gratis and for nothing, the equivalent of half a million tonnes of adjusted clean beet or, to put it another way, the production of 28,000 acres. But its far worse than this. Not only are they giving these crowns to British Sugar, but they are actually paying £2.8 million to transport the crowns from the farm to the factory.

The figures my friend has produced are hard to argue with – especially since they would appear to come from the files of British Sugar itself. However, his final conclusion is not one I could agree with. He suggested that British Sugar should be referred not just to the Monopolies and Mergers Commission (as I had suggested) but also to the Fraud Squad. Here, I fear, my friend misses the point rather seriously. British Sugar may have behaved greedily, and possibly immorally, but it has not behaved illegally. All of the regulations concerning crowns are laid down in black and white as part of the Interprofessional Agreement which governs the relationship between the Growers and British Sugar. The latter have, it is clear, kept strictly to this agreement. And who can blame them when the agreement is so massively weighted in their favour?

Of course in any other agricultural sector the solution would be simple; if the farmer was unhappy about the arrangement he would make sure that next year he sold to a different purchaser. But in the case of sugar beet growers there is no such choice because British Sugar is the only buyer. The enormous power of a monopoly doesn't just affect growers; it is also the reason why the voice on the end of the telephone didn't dare reveal his name. In the meanwhile the sugar beet growers of Britain will continue to donate £23 million of free sugar beet to British Sugar. But now, thanks to my anonymous friend, it is no longer a secret.

SUGAR BEET MONOPOLY
1997

Let's pretend it's time to select the cereal varieties you will grow next harvest. For most of us this is a pretty haphazard process; we may chat to a few reps and possibly wander round the local merchant's trials sites. Some of us study the NIAB (National Institute of Agricultural Botany) list and a few actually subscribe to organisations which run trials on behalf of their members. After we have made our choice we ring round the merchants and – just as we do when we are selling our grain – have a good old haggle which results in the price being shaved by a few quid a tonne. Buying seed is one of those events which mark the passing of the seasons – like the first asparagus from the garden or the last stubble to be ploughed.

But for some farmers these rules do not apply. I refer, of course, to sugar beet growers who have the exquisite pleasure of dealing with a monopoly. And monopolies, for those who have never met one, don't have to worry about customers. You may, it is true, select the varieties from the list the nice fieldsman (or in my case, fieldswoman) shows you. But after that brief moment of freedom you once again become a mere cog in the juggernaut which is known as British Sugar. Have you ever actually haggled about the price of sugar beet seed? Of course not. You pay the price British Sugar demands. No discussion, no arguments. Just shut up and write out the cheque. You are not even allowed to buy the seed direct from the breeders. British Sugar is a monopoly so they can make up their own rules.

When this year's beet seed was finally delivered I noticed to my surprise that I had been sent four varieties. This was strange since I had only ordered three. I phoned the factory and was told that since they had run out of one of my varieties they had sent me 'the next best one'. I protested but was told that this practice was permitted by the Interprofessional Agreement (IPA) and I had no right to complain. I would have been furious if my cereal seed merchant had made up my order of Riband wheat with a few tonnes of Brigand because it was 'the next best variety'. For most companies the customer is always right; for British Sugar the customer is always an inconvenience. But British Sugar is a monopoly so they can make up their own rules.

The day we started to drill our sugar beet I realised that I did not even know how much I was paying for the seed. I phoned my local factory at Bury St. Edmunds and was told that they did not know either. The price of beet seed is never decided until all the seed has actually been sold. What a wonderful way to sell seed – and what a terrible way to buy seed. But British Sugar is a monopoly so they can make up their own rules.

After two decades during which British Sugar rode roughshod over growers, things may be about to change. The NFU's sugar beet committee, under its tough chairman, Matt Twidale, is beginning to assert itself at this year's IPA discussions. Instead of meekly agreeing to everything British Sugar demands, Twidale is threatening to ask the Minister of Agriculture to appoint an arbitrator. He should, perhaps, go further and ask that British Sugar be referred to the

Monopolies and Mergers Commission. But in the meanwhile all growers should line up behind Matt Twidale. He's our last best chance.

Three years later, Matt Twidale is still locked in negotiations with British Sugar. I don't envy him his job. What could possibly be more boring than haggling over sugar beet? But, as they say in America, it's dirty work and somebody's got to do it.

FARMING AND THE ENVIRONMENT

VOTING GREEN?
1989

I almost voted Green the other day. Almost, but not quite. Like many other inhabitants of these islands, I was fed up with Mrs. Thatcher, turned off by Mr. Kinnock and disillusioned by Mr. Ashdown. In Cambridgeshire this left only the Greens. But on my way down to the village hall, I realised that for the first time in my life I was not going to vote at all.

Today I am deeply relieved because I now know what the Green policies really are. Two million people voted Green and I suspect that only a handful had actually gone to the trouble to read Green literature. They, like me, had a vague feeling that somehow the Greens cared. That somehow they were in favour of looking after the planet while the other parties didn't give a damn. In other words, the Greens were a cosy, friendly and, above all, sensible party.

For those who have not yet had the privilege of dipping into the Green manifesto, let me provide a few policies, some of which may surprise you as much as they did me. Cosy they may be, friendly they almost certainly are, but sensible they ain't.

As with most political tracts, it is first necessary to wade through swamps of invective. You find yourself immersed in 'chemical-laden crops and drugged animals'. All food which is not organic is automatically branded as being 'unhealthy' without any shred of evidence being offered to support this prejudice.

Their knowledge of agricultural practices is as vague as their solutions are irrelevant. Because of 'chemical fertilisers', we are informed, 'farmers fail to rotate their crops and specialise instead in massive monocultures'. In the real world the fashion for continuous wheat has long since passed away. Even on the 'prairies' (a shorthand term used by the Greens for everything that is worst about arable farming), I know of few farmers in Britain who practise monoculture. I was also surprised to learn that 'intensive production compacts the soil'. I had always imagined this to be caused by the weight of machinery rather than by the intensity of production. But the Greens (never people to let facts get in the way of propaganda) overlook these details.

When it comes to farming, their proposals would be ludicrous if they weren't also dangerous.

They will 'preserve genetic variety by scrapping seed patent rights and encouraging the use of local crop varieties and rare livestock breeds'. In other words there will be no plant breeding royalties and, as a result, the Greens will stop all plant breeding. So grind up your Hornet wheat and plant a bit of Squarehead Masters. Sell your Limousin and Holstein cattle and cross a Longhorn with a Chillingham. Send your Texels and Suffolks to the abattoir and buy a few Herdwicks and Jacobs for your flock. This is what the Greens call progress.

Not content with turning back the technological clock, the Greens will also destroy the structure of British agriculture which makes this country the envy of Europe. When it comes to land tenure they have a policy which will 'promote land ownership reform so that large farms can be broken up into smaller, often more productive, units'. A Green Party spokesman told me that 160 acres was considered to be the optimum size of a farm but that no firm decision had yet been taken. It was also unclear whether the large farms would be expropriated or bought at full market value.

The Greens will 'encourage extensive rather than intensive farming'. What form this encouragement will take is also far from clear. Whether they intend to use an (organic) carrot or an (inorganic) stick to encourage farmers has yet to be decided by the policy makers.

It comes as no surprise to learn that they will 'promote organic farming'. Apart from announcing that there will be grants for conversion to the organic system, they remain coy. Many tenants would like to know how the Greens propose to reduce their rents, which is a prerequisite of organic farming.

The Sudanese farmer watching helplessly while his sorghum is destroyed by billions of locusts will not be encouraged to hear that the Greens are going to 'help poor countries establish organic agricultural schemes'. He knows that the only effective organic method of killing a locust is to crush it between thumb and forefinger. The Indonesian rice farmer who, thanks to fertilisers and the Green Revolution, can now feed his own country, will not be overjoyed by a return to old methods and old varieties.

As part of the programme to encourage us to go organic, the Greens will 'drastically limit the use of nitrogen fertilisers'. The word drastic could, of course, mean many different things. It could mean a swingeing tax on nitrogen, or a rationing system or a simple quota. We await further developments with some trepidation.

The Green Party spokesman admitted that these measures would raise food prices by approximately 15 per cent.

When asked if the two million voters were aware of this particular consequence of Green policies, he assured me that they were. Green voters, I was informed, almost certainly knew that they would pay more for their food but they would not object since it was 'healthy' food.

At least on the wider stage of world politics the Greens are completely unambiguous.

'The Green Party believes that Britain must leave the Common Market and abandon the ludicrous pricing systems of the CAP.' They would also 'take Britain out of NATO' and 'would pursue immediate and unconditional British nuclear disarmament'.

The Green Party is, understandably, triumphant after attracting 15 per cent of the votes in the European election. It is, by any standards, a great achievement. Yet if I were a Green I would be uneasy about the future. There is, after all, a faint chance that some of those voters will one day read the Green manifesto.

In those dim and distant days Greens were a gentle lot who cared passionately about things – and lost no time in telling everyone. Today, however, there is a meaner and nastier bunch of Greens who smash their way on to farms and destroy the genetically modified trial sites. They do so in the name of democracy and feel that they are 'de-contaminating' the world. Why these people should be treated any differently to vandals who set fire to fields of wheat is a question I have yet to answer.

IN A PERFECT WORLD
1995

In a perfect world the sun would shine from Monday to Friday, rain would fall at weekends, oysters would cost 5p each and there would be no agricultural subsidies at all. But in the real world in which we live, the last point might just cause a few problems. It is true that a few farmers might survive on this unreal level playing field. Maybe some cereal growers in East Anglia could compete with Kansas and Saskatchewan in a subsidy-free world. But the environmental and social cost would be terrible. We too would have to farm 3,000 acres with a single man. We too would have to remove every surviving hedge to create 400-acre fields and we too would have to sacrifice everything on the altar of efficiency. But upland livestock farmers would have no such opportunity. There is nothing a Welsh sheep producer could possibly do which would enable him to compete with his New Zealand counterpart. Likewise a Scottish beef producer without subsidies would be eaten for breakfast by the Gauchos from Argentina. So if the farming population – let alone the landscape and wildlife – in Britain today are to remain at or near their present levels, we will need subsidies to protect us from Kansas wheat, Argentina's beef and New Zealand's lamb. Which is why – whether I like it or not – subsidies in one form or another will continue to be a fact of life. The question, therefore, is not whether subsidies are a good thing, but rather, how can they be paid to farmers in a form which keeps the public happy? All but the most blinkered and reactionary will agree that the old system of support prices, with its incentive to produce surpluses which must then be stored expensively and exported expensively, will not return. Ray McSharry's area payments may also be expensive but at least they managed to cut the link between subsidy and output. The snag is, of course, that they are highly visible. Visible not only when the IACS cheque comes through the letter-box, but also visible to the taxpaying public.

Which is why in the long run these area payments will not survive. Regardless of the facts, the public perceives setaside as simply paying farmers to do nothing. For farmers like me (big but not vast), my cheque will amount to £150,000 this year. For companies like Velcourt and CWS Farms, the cheque must be in the millions. Sooner or later (and probably sooner) the public and the politicians will get fed up with this system. So how will we ensure that money continues to flow into agriculture – and does so in a way which keeps the public happy and satisfied? The answer is simple; indeed, the system is actually being operated today. Ring up your nearest Countryside Commission office and you might get a shock. You will find, for example, that as part of the Countryside Stewardship Scheme you will be paid £300 per acre just for leaving a two-metre strip around your headlands. For want of a better word I call this a green subsidy. Now I know that most farmers' knuckles turn white and their faces turn red at the mention of the word green. But they should relax a bit. Greens and environmentalists aren't half as stupid or wicked as they have been painted by the farming media. In fact some of them are almost as intelligent as NFU Council members. Which is why it is crucial for farmers to stop heaping abuse on the Greens and instead join with them to present a united front to the politicians of Westminster and Brussels.

If – like most farmers – you'd sooner boil in hell than actually make common cause with a Green, then here is another reason why you should swallow your prejudices and change your mind.

One of these days British farmers are going to wake up to find that, in spite of John Gummer's victory two years ago, Brussels has decided to limit the amount of subsidy any single farm can receive. This will effectively cap all subsidies at around 250 arable acres, and will thus hit most UK farmers very hard indeed. But – and this is crucial – it will be impossible and wholly undesirable to cap a green subsidy. To do so would mean that an oak tree on a 2,000-acre Suffolk farm is somehow less environmentally benign than an oak tree on a hillside of a 20-acre Bavarian farm. Or that a hedge on the Lincolnshire Wolds is less valuable than a hedge in Devon. In fact, of course, the opposite is true. A tree on a Suffolk prairie should be encouraged with all the love, care and cash at our disposal. Likewise a stone curlew in arable Cambridgeshire is no less important than a vulture in the Pyrenees.

So Green subsidies make sense for four reasons:

They need cost (net) no more than today's subsidies;
They will be politically popular with the public and politicians;
They will unite Greens and farmers;
They will not discriminate against the UK.

In the meanwhile it will mean that Greens must talk to farmers and farmers must talk to Greens. We both might learn a thing or two.

> I wouldn't change a word of this article if I was publishing it today. Greens and farmers must talk to each other – which is why some of us got together and formed the Agricultural Reform Group (ARG). The CAP must be reformed, and one of the most important differences would be that subsidies should be given for environmental actions and not simply for planting more wheat.

COUNTRYSIDE STEWARDSHIP
1996

The letter from the Countryside Commission was clear. 'This is,' it said, 'a fraudulent claim and as yet it has not been decided what action to take.' Even a half-witted peasant like me knows that fraud is a criminal offence, so I took it very seriously. Almost a year earlier it had all been so different. After months of thought, I had taken a deep breath and had signed up for the Countryside Stewardship Scheme. The paperwork involved was enormous. Fortunately, however, the Countryside Commission were extremely helpful. Not only did they measure all the hedges and headlands, but they also filled out the forms for me.

Eventually, on 10 February 1995, I signed the document and became a happy and enthusiastic member of the Countryside Stewardship Scheme. In the following months we planted two miles of new hedges, coppiced and gapped up some old ones and spent £1,600 on a wildflower mix. We also received a large cheque which covered over half of our costs. All of this explained why last summer I was feeling even more self-righteous and self-satisfied than usual. And then that letter arrived. It appeared that I had claimed (and been paid) more money than a particular hedge had warranted. I was mystified, until it eventually dawned on me that the bureaucrat who had been checking on the work we had completed, must have been looking at the wrong hedge in the wrong field.

Anybody can make a mistake and on a large farm it is easy to go to the wrong field. But what made me incandescent with rage was the fact that instead of suggesting a chat to sort out this problem, he had simply written me a letter in which he accused me of fraud. Three weeks later, after two separate visits from the bureaucrat in question and his superiors from the regional office, I eventually received an apology. They admitted that I had not committed fraud and that they had, as I had suspected, been looking at the wrong hedge in the wrong field. But my problems were far from over. The regional boss had more news for me. I had, he announced, claimed too much money for the repair of yet another hedge and he would have to 'take the appropriate action'. This was strange because the claim form had actually been filled out for me by the Countryside Commission. I pointed this out to the regional boss and his reply took me completely by surprise. 'Ah yes,' he said, 'but your signature is at the bottom of the form and therefore you must take responsibility for the claim you have made.' This was more than I could stand and a few weeks later resigned from the Countryside Stewardship Scheme.

I hope my story is not typical because the scheme remains as good as it ever was. I was about to become more optimistic when I saw that MAFF had taken over the administration of the scheme from the Countryside Commission but last week I was told that the man who had once accused me of fraud had (no doubt because of his innate skill at handling farmers) himself been transferred to MAFF to continue his old job. So before any farmer decides to sign up to the Countryside Stewardship Scheme maybe he should think long and hard about my experiences. I wish him luck.

Perhaps I was a bit unfair and a bit hasty. Perhaps I let my anger influence my actions. The fact remains that the Countryside Stewardship Scheme remains an excellent idea. And, who knows, the bureaucrats who run it may even have changed their ways too. I hope so.

BADGERED AND BAFFLED
1997

I returned from holiday at the end of June to find some good news and some bad news. The good news was that we had just experienced the wettest month on record. On our thin south Cambridgeshire chalks we need water as often as an alcoholic needs booze. When I left home at the beginning of the month we were facing our second serious drought of the year; flag leaves on the wheat were curled and the sugar beet seemed to have stopped growing altogether. I returned to find an Amazonian farm on which everything was growing lushly.

The bad news I received was that a man from MAFF and a policeman had made an appointment to see me the following day. And the copper wasn't just the local village bobby. Oh no. He was, I was surprised to learn, a full-blown police inspector. What could I possibly have done to warrant such a formidable force? My guilty conscience (never far below the surface) swung into action. Murder? Manslaughter? Trafficking in home-grown cannabis? Grievous Bodily Harm? I didn't sleep too well that night, but perhaps that was caused by jet-lag as much as worry.

The Inspector arrived, sombre and very serious. Although he was wearing big black shoes, he had thoughtfully taken the trouble to wear a tweed jacket over his uniform so that my neighbours would not see me being interrogated by the Law. The MAFF man kept a respectful distance and didn't say much. 'We have received a complaint,' began the Inspector, 'and I am here to investigate.' I felt my knees wobble momentarily. After what felt like an age, he continued, 'It seems that you have been destroying badger setts which, as you know, is against the law.' The MAFF man nodded gravely.

I was amazed since I happen to love badgers. When we first found their setts some years ago we went to great lengths to inform the relevant naturalists of their location so that any interested parties would know all about them.

Half an hour later we were standing in a field of beans looking at a collapsed tunnel. The resident badgers, which had been living in a nearby spinney for the past ten years, had decided to vote with their feet and had dug a series of new setts 14 metres out into the field. Our sprayer, as it went along the tramline, had fallen in one of these tunnels.

Both the Inspector and the MAFF man took one look at the site and saw at once that no damage had been done to any badger and that the problem – if indeed it was a problem – had been caused by a normal agricultural operation. They were clearly surprised because they had been led to believe by my anonymous accuser that we had been engaged in a vicious and premeditated campaign to kill the no longer rare but certainly innocent, friendly and cuddly badgers.

The same accuser had apparently taken it upon himself to walk 14 metres through a standing crop of beans in order to check on the welfare of the badgers. Now I am not one who objects to walkers on my farm. On the contrary, I am only too happy for members of the public to walk along every track on the farm, even though they are not public footpaths. I am equally concerned that farmers respect and encourage wildlife wherever and however possible. But I must admit that I am very uneasy at the prospect of self-appointed wildlife guardians deciding to walk through my growing crops because they feel that there might possibly be a badger or skylark or fieldmouse being mistreated somewhere in the middle of the field.

Anyway, the result of my dealings with the police was simple and clear-cut. The MAFF man suggested – and I readily agreed – that we apply for a licence which will enable us to carry out normal agricultural operations on this field. In the meanwhile we will hope (maybe optimistically) that the badgers will once again return to their spinney and live happily ever after. My feelings about the anonymous complainant are, however, a lot less charitable. If he or she had had the courage, decency and politeness to telephone me instead of ringing the police, we would have sorted out the misunderstanding over a cup of tea. But like so many of today's villagers, he (she?) wants to have the pleasure of making trouble for farmers but lacks the guts to meet us face to face. As a result, an important and highly paid policeman wasted an entire afternoon standing in a field of soggy beans. Meanwhile the anonymous badger lover probably took time off from whingeing about wicked farmers and instead wrote to the local paper complaining about rising crime and the lack of policemen on the beat.

My problem was solved when the Ministry of Agriculture granted me a licence to plough the field – and with it the badger sett. But only a week ago I discovered another badger sett in the centre of a field of setaside. Once again, I have applied for a licence, and once again the badgers will have to return to the rough land beside the field if they want a peaceful life. I wouldn't feel so charitable towards them if I still had a dairy herd and my cattle were threatened with the tuberculosis which badgers carry.

GM CROPS vs THE MOB
1999

In the days since the Prince of Wales nailed his testament to the masthead of the *Daily Mail* I have stood aghast on the sidelines as the debate over GM food has raged ever more noisily. Like most farmers I am very confused by the prospect of genetic engineering and have – unlike my organic friends in the Soil Association – an open mind. I have not the slightest idea whether genetically modified crops will turn out to be either good or bad for the world. Which is why I cannot answer the Prince's fundamental question: do we need GM food in this country? I would, incidentally, also have great difficulty in knowing whether today we really need baked beans or – come to think of it – the *Daily Mail* itself. Hindsight is a wonderful gift which comes to all of us if we wait long enough. A century ago had anyone asked my great-grandparents if they needed a telephone they would have received a pretty dusty reply. Nobody needed the Wright Brothers' machine and at the time Fleming was beginning his research on moulds, few people would have been certain that the world needed penicillin. Thus I am not persuaded of the argument that proven need is a significant factor in technological advance.

Another strand in the Soil Association's argument which I find difficult to understand is their belief that the only people who will benefit from GM technology are the hated multinationals. I cannot see why any farmer – be he an East Anglian barley baron like myself, or an Ethiopian peasant growing sorghum – will buy GM seeds unless he wishes to. But to hear some of the GM haters talk you get the strong impression that Third World farmers will somehow have no option and will, as a result, be forced to give up their traditional practices and their traditional varieties because Monsanto orders them to do so. I know that on my own farm in Cambridgeshire I will only even contemplate using GM varieties in the future if I am reasonably certain that they do no damage to the environment and also will produce a crop which the consumer actually wants. On neither of these two points have I yet seen a single piece of evidence to either support or disprove the case for GM crops. And yet the organic lobby has apparently already made up its mind. I find this depressing but not in the least surprising.

The reason I am not surprised is that the opposition to GM crops is based on a philosophy. Some would even go as far as calling it a religion. Among these appears to be the Government's Chief Scientific Adviser, Sir Robert May, who believes the Soil Association is run by Ayatollahs. This is a trifle unfair because most of them are serious and passionate people united in an unshakeable and utterly rigid belief that their philosophy is right with a capital R. Unfortunately this certainty does mean that those of us who do not share their faith are dismissed as being at best short-sighted and at worst wicked.

It is perfectly true that today – at the early stages of the technology – most of the GM crops being tested are designed to resist particular herbicides, and thus increase the sales of these weedkillers. But the same technology could – and I hope will – be used to grow crops on saline or arid soil. To point this out is not – as the Prince of Wales suggests – emotional blackmail. Who knows? They may one day even engineer an elm tree which is resistant to Dutch Elm disease.

It is perfectly understandable for some people to announce they will not eat GM foods. This is called consumer choice and is a thoroughly good thing. It is also why detailed labelling showing GM ingredients should be mandatory. But the actual safety of GM foods is not – or should not be – at issue. The same cannot, however, be said for the effect of GM crops on the environment. Which is why the government is absolutely correct to forbid commercial plantings until tests have been carried out which satisfy all but the most intransigent opponents of GM technology. The only way we shall ever know what effects GM crops have on the environment is to continue the testing programme under strict controls.

To most people this would seem like good old-fashioned common sense. But to a small minority of fanatics even the carefully controlled trials must be stopped. These groups have now been so successful in destroying GM trial sites that CPB-Twyford, one of Britain's leading plant breeders, recently announced that they could no longer accept this level of wanton destruction and would, as a result, stop all work on GM crops. Chalk up another victory to mob rule.

Sixty years ago in Germany people in brown shirts smashed their way into private houses and burned the books they found on the shelves. They claimed that the ideas contained within these books would pervert the minds of Germany's young. Today in Britain people dressed in white coats are entering the private property of individual farmers and destroying the crops because they claim that the varieties being grown will cross-pollinate the plants of Britain. It is again depressing that not a single word of condemnation has been heard from the organic movement about these groups.

Until we know in detail what effects GM crops will have on the environment it would be folly for these varieties to be grown on a commercial basis. But unless trials are carried out we shall never know the answers. Which is why I am today inviting any bona-fide plant breeder to use my farm to carry out properly organised trials on GM crops. Only by doing this will I – and every other farmer – ever hope to know the answers. Until then I shall remain open-minded. Is it really asking too much of the organic movement – and the media – to be equally agnostic?

> I am still waiting for a GM company to agree to conduct trials on my farm, but things are looking promising. I am also encouraged by the fact that the militant eco-warriors who smash up GM crops appear to have alienated the general public who had previously been rather sympathetic to their aims – but not to their methods.

TRAVELS

MISSISSIPPI FARMERS
1994

The Peabody Hotel in Memphis is very grand indeed. It is also a little different. Every morning at 11am the doors of a lift open and, to the sound of a trumpet fanfare, out step a dozen ducks. They waddle across the crowded lobby on a red carpet to the ornate fountain where they spend each day.

But it isn't just the poultry which makes the Peabody Hotel an agricultural landmark. The Mississippi Delta, a vast crescent of some of the most fertile land in America, is traditionally supposed to start in the lobby of the Peabody Hotel and end on Catfish Row in Vicksburg, some 150 miles to the south.

Like the Fens, the Mississippi delta is a world of its own, untroubled by tourists, industry and – some would say – the twentieth century itself. Originally swampland, it never had the rich plantations and the beautiful houses which are so typical of the rest of the south. When the land was eventually drained, it was perfectly suited to a single crop. Cotton was king in the Delta. Today, 150 years later, cotton is still king, even if it is being threatened by soya beans, rice and catfish.

Two hundred miles north of New Orleans, the countryside changed. Beside the river, in the shadow of the levee, combines were finishing a field of soya beans and overhead an aircraft seeded wheat into an endless field. The farms were big and had equipment to match. Down on the Mississippi river bank a prison guard with a loaded revolver watched as prisoners from the state penitentiary cleared the undergrowth.

I knew I was in the Delta when the earth became black and the poverty visible. In Milestone, an almost derelict hamlet south of Tchula, I met Vicky (she didn't want to tell me her surname) at the petrol station. Although she did not look over 35, she was a grandmother several times over. In her arms she was cradling her latest grandson who was six weeks old.

'Everyone round here works on the plantations,' she told me. 'Of course, some planters are better than others. The difference is both in the wages and the way the farmer treats his workers. My daddy works on a plantation and in my opinion,' she added quickly, 'he is treated real poorly.' It emerged that he is paid the minimum wage of $3.35 an hour and receives a free house. 'It ain't a house really,' said Vicky, 'it's a shack.' I knew what she meant. The standard of rural housing in Mississippi is often reminiscent of a Third World country. The contrast between the plantation workers' houses and those of the farmers is particularly striking.

I wondered if there was a lot of unemployment. 'Yeah,' said Vicky, 'but not so much right now because it's cotton-picking time and there's work to do in the fields and in the gins. But wait 'til Christmas and there'll be lots without jobs. They ain't got no work 'til the spring when cultivating starts again.'

I later learned that farmworkers are inevitably laid off from December until March. For the first six weeks they receive unemployment pay of $120 a week and after that they must survive on their savings and on food stamps from the federal government. As if this were not humiliating enough, the farmworkers of the Mississippi Delta are not paid for overtime and receive no paid holidays.

My destination was Greenwood, Mississippi, the county seat of Leflore County, in the heart of the Delta. Greenwood describes itself, with typical American modesty, as 'The Cotton Capital of the World'. It is also home to one of the best restaurants in America. Lusco's is a shabby old converted grocery store on the wrong side of the tracks where, for four generations, Delta farmers have been eating stupendous steaks, catfish and gumbo soup. Lucky Delta farmers.

My first stop was the Soil Conservation Service which is an arm of the Department of Agriculture. James Johnson is the area conservationist responsible for ten counties. 'The Delta is a strange part of the world,' he told me, 'It's unique in the United States and we'd like to keep it that way if we can.' He went on to explain that, with an annual rainfall of 52 inches a year – almost all of which falls in the winter – soil erosion is a constant problem. James's advice to farmers has been not to plough in the autumn and to control weeds by chemical means rather than by cultivation. But however much sense this may make to a soil conservationist, it immediately runs foul of the environmental lobby which is today pressing for fewer chemicals to be used. 'There is a bit of a conflict here,' he admitted, 'and it's going to get worse.'

It is not just the terrain which makes the Delta unique. 'Delta farmers are a bit different too,' James told me. 'They tend to be old established families who use their own money rather than the bank's money. So the problem of being over-borrowed, which has affected many other parts of the United States, has been less common here. Of course,' he added quickly, 'there are bad farmers here in the Delta and they have had their difficulties, sure enough.'

Dwayne Bush is a good farmer who does not come from one of the old-money families. Indeed, his father began as a sharecropper in the 1930s and today three generations of Bushes run 6,000 acres at New Hope Plantation. Over a hamburger at Greenwood's second-best restaurant, the Crystal Club, Wayne explained that on his farm, as is usual in the Delta, cotton is still king. He is particularly lucky because most of his farm is the free-draining sandy loam which is perfect for cotton, rather than the heavy gumbo clay on which you can either grow rice or have catfish ponds.

Although Wayne was as courteous as all southerners inevitably are, he wasn't in a happy mood. Cotton picking was coming to an end and it was clear that the harvest was going to be bad. 'We'll be lucky to average 600lbs an acre,' he said, 'which is just over one bale (480lbs). Our average yield is around 1,100lbs per acre so you can see we're not going to have a good year.' The one bright spot on an otherwise gloomy horizon was that at least the price of cotton had risen sharply. 'It hasn't affected me,' said Wayne ruefully, 'because I sold forward at a lower price.' He had gone in for the Federal Support programme by setting aside 25 per cent of his cotton acreage. This at least guaranteed him an intervention price which was slightly above the market price.

When the conversation turned to soya beans, Wayne was even less cheerful. 'Beans are worse than cotton. The yield was 28 bushels compared to our average of 40 bushels, and the price is bad too.' I was wondering when to get the Kleenex out when he continued, 'I did, admittedly, sell half of mine forward for a very good price indeed, so at least I made one right decision.' He grinned.

Later that afternoon, as we travelled round the farm in one of the vast four wheel-drive pickup trucks so beloved of American farmers, I began to understand what he had been up against. We stopped at a field of soya beans. Instead of being up to my waist, the plants were a bare 12 inches high and had pitifully few pods.

'In a normal year,' continued Wayne, 'we double the crop beans by planting them after we've combined wheat in June. But this year it was so wet we had to wait until July and the beans never really had a chance. They should really have been harvested a month ago and we're now so far behind I don't suppose this field will see a combine before Christmas.' His gloom became even more profound when he admitted that the yield would not be enough even to cover the variable costs of the crop.

Back in the farmyard, some of Wayne's pessimism evaporated. 'Overall it isn't a total disaster. It certainly isn't the worst year I can ever remember, but yieldwise it is real bad and a lot of Delta farmers will make a loss this year. There's no doubt about that.'

After the recent rains the land was too wet for the cotton pickers and most of the farm staff were repairing machinery. I wondered how many men Wayne employed. 'We've got 13 people full-time,' he told me. I began to feel that Vicky had been exaggerating until Wayne explained what he meant by 'full-time'.

'Of course, we lay them off after we've finished picking cotton, but we take them back when there's work to do in the spring.' Unlike many of his neighbours, he also keeps his four best men throughout the winter, when they overhaul the machinery.

This year, however, Wayne will be short of one of his key men because a senior tractor driver had recently been killed in an unfortunate accident. 'He got drunk a couple of weeks ago,' explained Wayne, 'and just lay down in the middle of the road. A truck ran over him.'

Most of Wayne Bush's farm is irrigated. 'We started in 1954,' he explained, 'with an open furrow system of simple flood irrigation, and 11 years ago we bought our first centre pivot. Today we're on the edge of a whole new revolution called chemigation.'

Chemigation? That was a new word to me. Wayne explained. 'It's a brand new technique being developed for applying both chemicals and fertiliser through centre pivot irrigators.' He paused to let me digest this surprising concept. 'Of course its still in its very early stages, but if it actually works it will revolutionise farming because we'll only need tractors for cultivating. The rest could be done by a computer.' I assumed that the whole idea was still a long way from being a practical proposition, but I was wrong.

'It's the coming thing,' enthused Wayne. 'It's going to be as important to us as the arrival of the cotton picker just after the war. You remember the name. You'll hear more about it.' It made me keener than ever to buy a centre pivot irrigator.

'And chemicals aren't the only things either. They've already planted wheat, rice, milo and beans through these pivots. Of course you've got to have different nozzles, but it can be done. And they're now experimenting with sensors in the field which are linked to a computer and ensure that the irrigator turns itself on for the parts of the field which have a moisture deficit.' I began to think that I had stumbled into a science fiction dream world when Dwayne reminded me that these developments were still many years away from becoming realities. 'But you wait,' he said with relish, 'you just wait.'

I wondered if Delta farmers were coming under pressure from the green movement to convert to organic farming? His answer was definite. 'There's quite a movement in truck farming (market gardening) towards the organic system, but for a cotton farmer like me it would be quite impossible. This isn't just prejudice; it's a fact that we have to spray insecticides as many as ten times between planting and harvesting – not to mention four times with herbicides. If we were stopped using insecticides we'd be eaten alive by boll worms and nematodes.'

Apart from chemigation, had farming in the Delta changed much in the past decade, I wondered? 'Oh sure,' said Wayne. 'For a start labour has become much tighter, much more difficult to find. You must remember that traditionally all the workers here grew up on the plantation and never knew anything except farm work. They were born on the plantation, went to school in the plantation school, went to church in the plantation church and did their marketing in the plantation store – or commissary as it was called. They didn't pay cash for their purchases but it all went down in a book and at the end of the year, after the cotton was picked, the amount was deducted from their wages.'

Aven Whittington Jr. of Schlater (pronounced slaughter), Mississippi, is the sort of farmer James Johnson had in mind when he referred to the old-money families. His ancestors have dominated Mississippi politics and society for generations and today the family still runs a very large agricultural business. Aven Jr. himself looks after the 4,100 acres at Buckhorn Plantation, his cousin runs another farm, and his father, aged 71, supervises both of them.

Cotton is, not surprisingly, the main crop, but what makes Buckhorn Plantation rare is that it has its own – somewhat ramshackle – cotton gin. In the old days all farms operated their own gins but today the cost of machinery and labour makes this impractical. 'Ours is already paid for,' explained Aven, 'so we can gin our own cotton for less than a half what it would cost us if we took it to another gin in the area.'

The cotton bolls, looking like exactly what they are – cotton wool – are removed from the plants by cotton pickers which usually go over a field twice. Like combines, they store the crop in a tank which is tipped into a trailer beside the field. In the old days the trailers would be hauled straight to the local gin where the cotton seeds are removed from the boll. Today, however, a new technique is becoming widespread. The trailers tip into a large container which uses hydraulic rams to squeeze the cotton into a dense block, called a module. This machine, which amounts to a stationary baler, enables cotton to be transported from the field to the gin with far fewer trailers than used to be necessary.

But in spite of new techniques, cotton is still a labour-intensive crop to grow. 'So,' said Aven Jr., 'there is no possibility of farming cotton like a Midwest grain farmer does his wheat – with a son and maybe a hired man. It also explains why Delta farms are relatively big, with an average size of 1,000 acres.'

When the subject of labour came up, Aven Jr. had strong feelings. 'We keep on our 13 full-time workers throughout the year because we need good labour badly,' he told me. Then there was a pause. 'Well, we use the unemployment system by laying some of them off for a week at time so that they can claim unemployment benefits. Then we take that group back and lay off the other group.'

I was trying to digest this information when he continued. 'Ten years ago there wasn't a labour problem. We could go into Schlater and pick someone up. In those days anyone could do the sort of jobs we needed. But today the machinery has got bigger and more complicated so I'm really looking for a man with a high school education and these people are not easy to find. That's where my biggest problem is today.'

I wondered what difference there had been since Aven had been a child growing up on Buckhorn Plantation? 'Well for a start there's been the civil rights movement and, as a result, integration. It has all helped us a bit because it's given the labour force some self-respect which they lacked before. There's now more of an incentive for them to go and get a good education whereas there wasn't much before. We've now got a black congressman in this District who'll probably go on getting himself elected for as long as he wants to unless (he laughs) he's caught sleeping with a live boy or a dead girl.'

James Townes is a large man. As well as being a farmer, he is also County Fire Commissioner, which explains the mountain of equipment and red flashing lights which he carries in his jeep. He was smoking a large Jamaican cigar when we met on the lawn outside his house by the banks of the Talahatchee River. A few minutes earlier I had noticed a sign by the roadside which announced 'Palo Alto Plantation 1831'. James explained that his great-great grandfather had bought the plantation from the United States government in 1831. In those days the 45,000-acre tract of land was nothing more than a swamp which the Choctaw Indians had recently vacated. Since then, the plantation has remained in the Townes family, though today it amounts to a mere 4,500 acres on which James grows cotton and soya beans.

Over lunch served by his cook, he told me a bit about how times had changed for the plantation owners in the Delta. The original farming operation had, of course, been based on slavery, but after the Civil War the slaves were freed and many of them moved away to look for other jobs. These families were replaced with sharecroppers who were effectively tenants on the estate. James's predecessors provided the land and seeds while the sharecropper provided his family's labour. The proceeds were split equally between landlord and sharecropper. However, no cash actually changed hands since the sharecropper's reward was in the form of credit at the plantation's store.

By the late 1950s sharecropping had disappeared in Mississippi, as had the manual picking of cotton. But even while James was growing up, the Palo Alto estate was home to 80 families of sharecroppers; today it employs eight full-time men. The definition of 'full-time' did need a bit of interpretation since, in Mississippi farmspeak, it means a guaranteed working week of three days.

I was becoming slightly confused by the distinction between a plantation and a farm. The answer, according to James, was social rather than agricultural. 'Take a look at this yard,' he said, pointing to 15 acres of lawn and ancient pecan trees. 'I employ three yardmen full time to keep the place looking like it does. Now go and speak to a farmer and see how many people he has in his yard. He won't have anyone.'

Thus a plantation in Mississippi is similar to a landed estate in England. But, like so many English landed estates, the Palo Alto plantation today is also a working farm which has to produce a profit to survive.

We got into James's Jeep, with its licence plate FIRE, and I clambered over the radio equipment needed to keep in touch with the volunteer foreman in his county. 'Sorry about the mess,' he said, 'but I need it all for my job. I also keep a loaded .38 revolver in here just in case.' The threat, I was relieved to learn, came from wild dogs and coyotes. We drove through deep mud to a field where the plantation's three cotton pickers were in action. Passing a particularly dense wood, I asked what forms of wildlife lived there. 'Nothing much,' said James. 'Some white-tailed deer, maybe a wild turkey or two and four pairs of alligators.' Alligators this far north? 'Yes,' explained James patiently. 'Fifteen years ago we put them into the forest to eat all the beaver which had been blocking our drains. I guess they've done a good job because there are far fewer beavers these days.'

Out in the field the machinery was struggling. Two men were standing in a cotton trailer treading down the loads which the pickers tipped in. They were friendly, cheerful and talkative, in contrast to most of the other farmworkers I had met in the Delta.

My next appointment was with a man who was very different from James Townes. Hugh Arant, who farms 4,500 acres of gumbo clay at Ruleville, is probably the best-known farmer in the state because, until this year, he had been President of the Mississippi Farm Bureau for 16 years. We sat in the kitchen eating pecan nut cake covered in cream which his wife had just baked and Hugh explained why his farm was different from James Townes' plantation. 'This gumbo clay is about as heavy soil as you can find anywhere in the United States. They say that if you stick to it when it's dry, it'll stick to you when it's wet.' A glance at his mud-caked shoes showed what he meant.

In the old days the Arants used to grow cotton, but it was never very satisfactory. The drainage was bad and cotton doesn't like to have its roots wet. In the 1950s farmers like Hugh realised that there was an alternative crop to cotton which actually liked strong land: rice.

'It's been the best thing which ever happened to farmers like us,' said Hugh. 'Today rice is our main crop, and we grow quite a few soya beans too. But these aren't our only crops, because 15 years ago another revolution took place down here in Mississippi. We discovered a new enterprise which today has overtaken beans to become the second biggest crop in the state, after cotton: catfish.'

The evening sun was setting when Hugh took me out to some of his farm's

365 acres of catfish ponds and explained how the industry worked. 'We were looking for things to do with this heavy old gumbo and one of the farmers near here got the idea of catfish. They're good to eat and they live naturally in all the rivers round here but nobody had really tried farming them. We set up a cooperative called Delta Pride to process the fish and before long we had 170 farmer members. But the secret to our success was in the marketing of this new product. Eventually our big breakthrough came when we got the US Army to feed catfish to the soldiers. The Mississippi Congressmen and Senators invited the Secretary of the Army to lunch and gave him catfish up in Washington. Pretty soon after, the Army began to feed catfish to the troops and we were in business in a big way.' Today there are 90,000 acres of catfish ponds in Mississippi alone.

The baby catfish starts life as a fry, but when it gets to be an inch long it becomes a fingerling. Hugh Arant buys fingerlings about six inches long and stocks his ponds with them every March. By the following November the catfish weigh over 1.5 pounds and are ready for market. Thus the cashflow is good, and the profits aren't bad either. It costs around 50 cents per pound to produce a catfish which is worth 70 cents.

The Arants' catfish ponds contain 2,000–3,000 fish per acre, which is a moderate stocking rate. Some of his neighbours manage to keep up to 5,000 fish per acre, but the problems of disease become a great deal more severe.

As we stood on the edge of a pond and watched cormorants and herons waiting for their supper, Hugh explained how you convert arable land into catfish ponds. 'First of all you need to have good heavy clay. Then you take an 80-acre block and dig four lakes, each four feet deep, with a borehole in the centre. The whole operation now costs about $3,000 (£1,500) an acre.'

I wondered how Hugh had adapted to fish farming? 'Well,' he said in that slow drawl which makes a Mississippi accent so seductive, 'I'm really a cotton farmer at heart. I like catfish but the problem is that you can't ever see them grow. It's nice to see things grow isn't it?' I agreed.

Sunset over the Mississippi Delta had become spectacular as Hugh Arant said goodbye. His wife came rushing out of the house with a small parcel. 'Here,' she said, 'it might be difficult for you to take catfish back home, but at least you can have some real Mississippi rice. We harvested it last month and hulled it last week.'

I drove west towards the setting sun and the Mississippi River. That night, over a supper of fried catfish, I found that I was confused by the Delta. The land is magnificent, the farming excellent and the hospitality unrivalled. Yet I kept remembering Vicky and wondering if her six-week-old grandchild would grow up to work on a plantation. For his sake, I hoped not.

FIELD OF DREAMS – OR LEVEL PLAYING FIELD?
1992

Once upon a time farmers were a provincial bunch of people, interested only in what went on in their town or county. Only the exceptional ones knew or cared about what happened in the State capital, and you had to be positively eccentric to have an interest in world affairs.

Today, however, the opposite is true. Listen to a bunch of farmers these days and the subject of GATT will take precedence over the price of hogs or the college football results. And yet a casual visitor from Mars might be forgiven for thinking that the farming fraternity these days is obsessed with football. On both sides of the Atlantic, from North Platte, Nebraska, to Northallerton, Yorkshire, a phrase keeps cropping up in farmers' conversations. 'If only,' says one of the participants, 'we had a level playing field,' and the others nod vigorously. The odd thing is that the same conversation takes place in Europe and America alike. Both are resolutely convinced that the playing field is not level, and that it has been tipped against them. Will the level playing field, they wonder, always be a field of dreams?

As the GATT talks stumble on, and as the pressure to reform the CAP grows daily, it is beginning to look as if the playing field may one day be level – or at least less unlevel than it is today. As an English wheat farmer who has enjoyed playing football downhill on the playing field known as the Common Agricultural Policy, I was beginning to feel a mite uneasy about my chances of survival on a level field.

Throughout the 1970s raising wheat in England had been a licence to print money. Not only did the prices rise each year as the farm organisations warned of widespread bankruptcy throughout Europe but, thanks to better varieties, better techniques and better fungicides, our yields had doubled. In the early 1970s I used to aim at 80 bushels an acre, and be moderately happy if I managed 60. By the end of the decade I was shooting for 160 bushels and was regularly managing 120. My only problem was knowing what to do with the money. It was, however, a problem I never found too arduous, since if all else failed, I could (and did) always buy more land.

The results on my farm were mirrored throughout northern Europe. As a result Britain went into the 1970s as a substantial importer of wheat and came out of the decade as one of the world's largest exporters. This brought us into sharp conflict with the wheat farmers of North America, whose markets we were stealing with our heavily subsidised exports. The playing field, never very level at the best of times, was beginning to tip sharply.

Good times never last, and mine were no exception. By the mid-1980s both the politicians and the public woke up to the fact that they were paying me vast subsidies to produce a crop which was in surplus already, were paying more money to store the stuff, and yet another subsidy to export the stuff into a world which was already awash with wheat. By this time American farmers had also begun to feel the hard edge of our presence in the export market-place.

Today, some five years later, European farmers are still on the downslope as our subsidies shrink, our prices fall but our costs remain resolutely high. And as if this is not bad enough, the EC finds itself under growing pressure at the GATT negotiations from America and the Cairns Group. I have no doubt that by the time this article appears there will have been an agreement and European farmers like me will find that the slope on the playing field has become rather less steep.

To find out what the future holds for high cost wheat farmers like me, I decided to check out the opposing team, to gauge their strengths and weaknesses and – above all – to see how I would fare if the day ever came when the playing field was completely level.

The strongest team of wheat farmers in the world comes from Kansas, so it seemed the obvious place to head for. I left my home village of Thriplow, some 50 miles north of London in the flat landscape of East Anglia and flew west. That afternoon I arrived in Kansas City in time for a late lunch at my favourite restaurant in the world. A few hours later, full of barbecued ribs from Arthur Bryant's, I was on my way west again to the airport at Hays, Kansas.

A tall, man with thinning hair, a checked shirt and blue jeans was waiting for me. 'Hi,' he said shyly, 'I'm Alan States.' Two hours later I was being welcomed by his wife, Carolyn, at their long, low bungalow outside the small town of Logan. That night I slept in a waterbed and woke feeling as if I had swum rather than flown the Atlantic.

Alan States and I are, in many ways, extraordinarily similar farmers. We are of similar ages, both married with three children and both active outside the farm. Alan is active in the Canola Growers Association and – to my surprise, also owns a small bank in the neighboring town of Palco. I spend a lot of time travelling the world, writing and broadcasting about agriculture.

Alan's office is not in his house but instead occupies an old storefront on the shabby and almost derelict Main Street of Logan next to the Trojan Café where a few farmers were drinking coffee. I was pleased to see that he, like me, was addicted to computers. Within minutes we were deep inside a spreadsheet looking at his costs, which he records in great detail.

From a purely agricultural standpoint, our farms are strangely similar. He has 5,500 acres of land in eastern Kansas while last harvest I farmed 3,000 acres in eastern England. His land is, however, summer-fallowed, so he cultivates around 2,750 acres every year. We both concentrate on wheat, although I grow sugar-beets, rapeseed, peas and beans. Alan also grows some rapeseed, or Canola as it is called in north America, as well as sorghum and sunflowers.

While the crops we grow may be similar, the methods we use are worlds apart. Ever since the early 1970s I have used an extremely intensive, and hence expensive, system for growing wheat.

Table One (page 62) shows what happens to a typical field of wheat on my farm last year. Compared to previous harvests, we used fewer chemicals than normal, but compared to Alan States, we went hog-crazy.

TABLE ONE

Date	Product applied	Product type
9/27	Haven	Seed
10/26	Cypermethrin	Insecticide
11/7	Javelin	Herbicide
11/7	Hytane	Herbicide
12/4	0-24-24	Fertiliser
3/14	Urea	Fertiliser
4/8	Urea	Fertiliser
4/10	Cycocel	Growth regulator
4/19	Urea	Fertiliser
6/17	Dorin	Fungicide
6/17	Mainstay	Fungicide
7/3	Spinnaker	Fungicide
7/3	Tern	Fungicide
7/11	Aphox	Insecticide

At the end of the year, when the combines came into the field, the yield was a typical 126 bushels per acre. Our worst field barely managed 100 bushels and our best field exceeded 150 bushels. It is worth bearing in mind that the wheat both Alan and I are producing is not strictly comparable. Haven, the wheat I was growing is really only good for feeding hogs, and is a far lower quality than Alan's Hard Red Winter.

Nevertheless, a combination of a high price for my wheat of around $5.11 per bushel, added to a yield of 124 bushels, meant that my gross income came to a massive $633.64 per acre compared to Alan's $143.50. No wonder that all Kansas farmers think I have taken leave of my senses when I tell them about my farm.

After we had compared our harvest results we began the process I had been dreading – the examination of our fixed costs. It began well for me because Alan hires a custom operator to cut his wheat, costing him nearly $21 per acre, while I use my own three large Claas combines. So for a few seconds it looked as if I was going to win big. But then the bad news started, and continued until we broke for a sandwich.

Alan employs one hired man, and then only because he spends so much time on Canola Association business. Like most Kansas farmers of his size, he and his family look after the farm and only use a bit of help at harvest. I, on the other hand, employ six men, which explains why it costs me nearly $57 per acre for labour compared to Alan's $3. My machinery depreciation runs at a whopping $70.50, almost ten times higher than Alan's. This is because in the glory days of the 1980s I used to buy new machinery, take the write-offs and pay no tax. Now my folly is catching up with me.

When it comes to rent, Alan owns most of his land and share-crops the remainder while I pay a cash rent to my landlord. For the purposes of comparison we assumed that both of us paid a cash rent, and here again mine was two-and-a-half times more than Alan's. There is, I suppose, some argument to suggest that the Kansas rent is high compared to Cambridgeshire, since our outputs are so far apart.

Anyway, at the end of the day we had come up with the final figures (Table Two). They showed, to nobody's surprise, that I was still making a good profit growing wheat.

TABLE TWO

	KANSAS	CAMBRIDGE
Wheat price ($/bu)	3.50	5.11
Yield (bu/ac)	41.00	124.00
GROSS INCOME/ACRE	143.50	633.64
$/acre	$/acre	
Seed	5.00	32.70
Fertiliser	17.10	48.50
Sprays	12.30	54.40
VARIABLE COSTS	34.40	135.60
Paid labour	5.19	56.80
Custom combine	20.76	0.00
Equipment repairs	8.66	13.80
Electricity/energy	0.84	5.20
Office expenses	0.12	1.70
Professional fees	0.22	5.20
Property taxes	1.80	3.40
Depreciation	7.21	70.50
Rent	30.58	81.70
Bank interest	1.65	4.30
Insurance	2.16	12.00
Travel	0.42	1.70
Telephone	0.85	3.40
Fuel	5.00	15.50
Sundry	1.15	18.90
TOTAL FIXED COSTS/ACRE	86.61	294.10
GROSS MARGIN	109.10	498.04
Variable cost/bushel	0.84	1.09
Fixed Cost/bushel	2.11	2.37
BREAKEVEN PRICE/BUSHEL	$2.95	$3.47

Computed at an exchange rate of: £1 = $1.72

The comparison between a dry land Kansas acre and a Cambridgeshire acre is more than a bit misleading. For a Kansas farmer to enjoy my standard of living, he would have to farm 15,000 acres of land. In other words, one

Cambridgeshire acre generates as much profit as five Kansas acres. So in the short run Alan had cause to envy me and my apparent prosperity. But when we looked into the future and wondered what would happen on that mythical level playing field, the picture became very different indeed.

Alan's production costs are so much lower than mine that he would be able to survive (even if he did not flourish) with wheat at $3.00 per bushel. At that price I would have long since gone belly up. I need a price of at least $3.50 to survive, and even that would be a struggle I would find exceedingly unpleasant.

After two days of talking and two nights on board a waterbed I came to an unpleasant and disturbing conclusion. Neither of us are typical farmers, and our figures represented no more and no less than the figures of two individual farms. But if the day ever comes when GATT levels the playing field, I have no doubt that Alan States and the best of his fellow Kansas farmers will still be growing wheat long after Oliver Walston is just a distant memory. Of course by then my land will be being farmed by a newer, meaner, leaner English farmer whose rents will be lower, whose costs will be lower and who will be able to give Alan a real run for his money on the good old level playing field.

> There never will be a level playing field, even though the whole objective of the World Trade Organisation is to create one. But what will happen is that we shall all move very slowly away from the ludicrously high subsidies which both Alan States and I both enjoyed in the 1970s and 1980s. Alan and I will both continue to produce wheat in our very different ways, both of us will (like all farmers) be certain that the other is receiving unfair subsidies, and both of us will survive and even prosper.

WALSTON IN NORWAY
1990

Norway takes a bit of getting used to. The dial buttons on telephones are back to front, the salt shakers have three holes while the pepper shakers have only one, and a Norwegian mile is ten kilometres (six miles) long. So when a farmer said to me, 'I'm eight miles from town, it's a long way,' I assumed he rode a bike. Forty-eight miles later, I knew better.

The north of Norway is particularly different. For a start, the sun does not set for ten weeks during the summer. But, even more significant, northern Norway is where the farming stops. Thanks to the Gulf Stream, which keeps the climate mild, it is possible to farm further north in Norway than anywhere else in the world.

My journey started half way up the coastline in the small fishing port of Sandnessjoen. Here, 60 miles south of the Arctic Circle, was the limit of cereal growing. I was met by the local Agricultural Extension Officer called Marit Dyrhaug. Like most Norwegians – but unlike most Extension Officers – she was blonde. Over a coffee she told me a bit about agriculture in her district of Helgeland. With an annual rainfall of 1600mm per year, it is one of the wettest places in Europe. No wonder both the grass and the cows looked so good.

Dairying had not always been so important, explained Marit. 'Traditionally the farms in this part of Norway were mixed, and were run by the wife. The husband was a fisherman who helped with harvest but spent most of the year at sea. This system flourished until 20 years ago, but today, as in the rest of Europe, farms have grown bigger and more specialised. As a result there are no farmer/fishermen left any longer.'

Was it true that there were arable farmers in Helgeland, I wondered? 'Yes,' replied Marit, 'but they're so few, I can count them on the fingers of one hand. The conditions just aren't right for cereals here, but some individualists do grow barley for grain instead of the normal green fodder.'

The next day I went in search of one of these individualists and took a ferry over to Mindland, a small island famous for its potatoes. The total population is a mere 150, all of whom appear to live in white wooden farmhouses next to red barns. As I walked up the jetty, a man in a vivid green shirt waved at me. 'Welcome,' he said.

Viggo Skulstad is 42. With a pink face and the wispy remains of blond hair, he looks, sounds and acts like a typical Norwegian farmer. But appearances are deceptive. In the first place he farms a total of 32 hectares, which makes him a big farmer by Norwegian standards – and enormous compared to his neighbours on the island. Secondly he raises 27 hectares of barley which must make him the most northerly serious arable farmer in the world. (I did, as it happens, discover a pig farmer further north who grew four hectares of barley, but he admitted he did so as a hobby). Thirdly, inside his spotless farmhouse there are signs of a foreign influence. The coffee I was offered was Turkish, and his wife spoke to the two fair-haired boys in a language I didn't immediately recognise. It was Hebrew. Mrs. Skulstad is an Israeli of Bulgarian and Greek descent, which would not be at all surprising in most of Europe, but seemed slightly exotic among the pale Nordic Lutherans of sub-Arctic Norway.

As well as his barley, Skulstad has four hectares of potatoes and 19 sows. But cereals are what he cares about. 'My normal yield is around 3 tonnes per hectare,' he told me, ' but in this part of the world yields vary drastically from year to year. Last year, for example, I managed to harvest only 0.6 tonnes per hectare.' He grinned.

At this point an old man ambled up to where we were talking outside the barn. It was Skulstad's father who, like so many Norwegians of his generation, had spent the war years in England.

The old man explained, 'I started growing barley back in 1957. There wasn't really a good reason. I suppose I was just interested in the idea. Of course, my neighbours thought I was mad. They probably still do because none of them grow barley today. A few, it's true, grow it as a green fodder crop for their cattle, but certainly not for the grain.'

Viggo interrupted. 'I've nothing against livestock myself. In fact I used to have sheep but I gave them up a few years ago because all the work came just at the time I was busiest. In the autumn when I was meant to be harvesting barley and potatoes, I also had to take my boat over to the mainland and collect my flock from the mountains where they had spent the summer. This took a lot of time and effort and so I eventually decided to concentrate on barley and potatoes.'

His father continued. 'Those early days weren't a lot of fun. We had an old bagger combine and all the sacks of grain had to go over to the mainland to be dried before it could be sold. We loaded them into the boats and they invariably got soaked, so by the time they arrived at the merchant's the corn was even wetter than when it left the farm. The deduction for drying charges wasn't much less than the barley was worth.' I was beginning to understand why not many other people try to grow cereals this far north.

Did they still have a problem with moisture? 'Compared to you English farmers, I suppose we do,' said Viggo. 'We normally combine barley when it is around 28 per cent and have to dry it down to 17 per cent. But there have been times when I've cut it at 50 per cent. I don't do this very often, mind you. Nor do I combine barley on 24 December as I did back in 1983. That harvest had been so wet we abandoned an entire field. But then one cold day before Christmas I suddenly realised that the frost had penetrated the ground deep enough to allow the combine to work. We went out and managed to salvage three tonnes. It was lousy quality but at least it brought in a bit of cash.'

Viggo was warming to his task of telling a soft Brit like me about farming in the raw. 'In this part of the world,' he continued, 'you've got to do field work when the weather permits. There's no point in farming by the calendar here because the weather is so unreliable. One of the lucky things about having so little darkness in the summer is that at least you can do your work round the clock if necessary.'

Viggo Skulstad normally starts drilling barley in the first week of May. This year conditions were so good he managed to begin on 18 April, at which time he was optimistic about harvest. Since then, I learned, northern Norway has had a cold, wet summer. The day I visited him his optimism had evaporated completely. 'Far from being advanced,' he told me sadly, 'the barley is now as backwards as I have ever seen it in late July. But this isn't as serious as it would be in England because of our peculiar climate. You see, the sun shines for 24 hours a day at this time of year, which enables plants to grow very fast indeed.'

Did he ever have to spray the barley against disease? Viggo thought for a moment and said, 'Well I did once, but that's the only time I can remember. The fact is that our temperatures are so low here that the diseases are never a problem. This makes me happy because too many sprays are used today and I'm very conscious of the problems of pollution. Not just here in Norway, but throughout the world.'

Did this mean, I wondered, that Viggo Skulstad was organic? 'Oh no, not at all. I use chemicals when I have to and I have no objections to them. I just don't like using more than I need.'

Once harvested and dried, his barley is sold to a mill on the mainland. But to reach it involves a 60-mile journey and two ferries. 'We produce around 90 tonnes in a good year,' Viggo told me, 'which I cart with my tractor and trailer. It's a long day; I start at 8.30am after I have fed the pigs and manage to get back late in the evening. But the compensation is that, in addition to the £275 per tonne I receive for the barley, there is a fairly generous transport allowance too.' I discovered that the price of barley – and all other agricultural products – rises the further north you are.

Viggo Skulstad offered to take me back in his motorboat so I didn't have to wait for the ferry. As we crossed the fjord he noticed that a man was working on a series of floating pontoons. We landed on the raft where I was greeted by another sort of farmer – a salmon farmer.

Harliev Pedersen, a stocky, florid 42-year-old blond, was feeding his salmon when I stepped onto the rafts to which were attached two rows of salmon pens. As in an intensive piggery, automatic feeders (powered by a solar panel) were dispensing pellets at regular intervals. But, unlike a piggery, feeding was silent. Instead of being accompanied by deafening squeals, the water appeared to come to life as 3,500 large salmon fought for food. 'We buy them as two year olds weighing 100 grams,' explained Pedersen, 'and sell them 18 months later at around 4kgs when they are worth about £11.50. This may sound a lot but you should remember that when the fish are big as they are now it's costing us £1,800 per day just to feed them.'

Had there been any problems with pollution as in some of the Scottish lochs, I wondered? 'No, not really,' he replied somewhat predictably, 'but we have to move the farm around every three years. We'll sell our fish at the end of this year and will move the farm before we buy in the smelts during May. I suppose we shall shift it at least one kilometre so that the water and surroundings are fresh.'

On my way back to town I noticed a sight I had never seen before. A team was at work making big bale silage, but instead of baling the wilted swath, the baler was being fed directly by a forage harvester. The farmer, a young man called Anders Svarstad, explained that this was a new system. 'I belong to a group of five farmers and we started making big bale silage last year. We liked it so much that this year we've invested in all new machinery and so far it has worked well. It certainly speeds up the job a lot.'

But what about the dry matter content, I asked? 'Well, we've never been able to wilt in these parts anyway,' he explained, 'because it's much too wet. That's why the dry matter is never more than 20 per cent.'

Until clamp silage appeared ten years ago, all grass was made into hay, using the characteristic Norwegian Hesje method. The rainfall is too high to enable the hay to dry in the swath so instead it is laid over wires strung between posts. Three weeks later (two weeks if the weather is good) the hay is taken down from the frame and carted to the barn. In some parts of northern Norway, however, the winds are so strong that the hay has to be netted to stop it blowing away.

Halfdan Berg was one of the rare farmers who still used the old system. 'I've done it this way all my life and I don't want to change now that I'm 70,' he told me as he rested on his pitchfork. 'I know it's a lot of work but I reckon the hay is better for my sheep and,' he added with a twinkle in his eye, 'have you ever tried feeding silage when it's frozen solid?'

Halfdan was an interesting man. 'I've tried barley, and I've even tried wheat up here. We grew them in the war during the German occupation, but we had to do it secretly on very small fields because it was forbidden and the penalties were very harsh. But of course in those days the whole of my family helped with the harvest and so there wasn't a problem with labour. Today I'm on my own with my thirty sheep.'

Halfdan's flock consisted of native Dale ewes which he runs on 80dekar (one dekar = 1/10th hectare). He claimed that the lambing percentage was 200 per cent but, like flockmasters the world over, he was clearly a trifle optimistic. 'We're still suffering enormously from Chernobyl,' he told me. 'In fact the radioactivity this year is no better than it was last year. The odd thing is that it is confined to certain areas rather than the whole region.' He was amazed when I confirmed that we had the same phenomenon in Britain.

The island of Heroy is well-known to ornithologists as it provides the resting ground for Barnacle Geese flying between Scotland and Spitsbergen. It is a beautiful spot, full of old red-painted buildings with traditional grass roofs. To make these uniquely Norwegian coverings, birch bark used to be stretched across the joists and soil spread on top, out of which grew a layer of turf which was both weatherproof and attractive.

One of Heroy's inhabitants was Helge Lenning. His father, who had been one of the farmer/fishermen I had heard about, used to milk two cows on two hectares. Today the farm has expanded to 18 hectares and the same number of Norwegian Red cows which Helge milks with his son, Torigeir. The herd average of 7,000 litres was exceptional for the region, but the price of 40p per litre was standard for this part of Norway.

'I'd like to have a couple more cows because we've got enough land,' Helge told me, 'but we have a quota in Norway just like you and so it's impossible.' Instead, he grows two hectares of potatoes which he sells to the locals. Planted in mid-May, they are lifted four months later and yield around 35 tonnes per hectare. 'That's what you can do with long days like we have here,' explained Torigeir.

Did they have any problems with silage effluent from the clamp, I wondered? Helge grinned. 'No. It just runs straight into the sea as it's always done. The sea has always been the dumping ground for all the people in this part of Norway, but it won't for much longer because the government is tightening up on all these things.' He went on to explain that the government had also introduced an environmental tax on fertiliser which had raised the price of a typical compound by around six per cent.

'Life's getting tougher for farmers round here,' said Helge. 'In my Dad's day a family could manage on two cows and two hectares, but today you need almost ten times that amount of land and most farmers don't have it. Take my neighbour, for example. He's started to drive a bus as a part-time job because he can't make ends meet with prices as low as they are.'

It sounded like yet another farmer's whinge to me. Forty pence per litre for milk and almost £300 a tonne for barley could not possibly be called 'low prices'. Then I remembered I had paid £3.50 for a half litre of beer in the hotel the night before. The cost of living in Norway is mind-boggling.

It was time to move north again and cross the Arctic Circle in search of farming at the very limit. Stupendous scenery of mountains and fjords gradually gave way to a treeless arctic plain where reindeer grazed and bright orange cloudberries grew in the endless bogs. Beside the sea fishermen were drying their catch on frames bigger than houses, and beside the road Lapps (resembling a cross between a gypsy and a Scottish crofter) sold souvenirs to passing tourists.

The North Cape is one of those magical spots on the globe where people go just to say they have been there. At 71 degrees north, it is the most northerly point in Europe, standing on a cliff overlooking the grey Arctic Sea. Thirty miles south of the North Cape is the small village of Olderfjord where I found Georg Hansen forking hay into his barn. 'I'm a fisherman more than a farmer,' he told me, 'but my father died last week and I've come home to help my mother.'

It was a sunny day and 50 sheep were peacefully grazing on a nearby hillside. Was Georg going to take over the farm? 'No, I don't think so. I'll probably sell the flock next year when they've lambed. I don't really like this job and there's not enough money in it any more to support me and my family.'

'I suppose,' he continued, 'I've been born a bit too late. In my grandfather's day you could be part fisherman and part farmer but my father managed to be only a farmer. I wouldn't want to do today what he did when he was young. Then he had two cows and two horses. The horses were used to drag telephone poles over the hills for the post office and the cows produced the milk which was sold.'

Today Georg Hansen's farm was, inevitably, specialised. The sheep lamb in April and the lambs are sold in September at around 15 kilos when they fetch £45. The ewes are sheared in October when they come in for the winter after a brief three months in the open air.

Gunnar Olaussen appeared over the top of the hill on his Valmet tractor hauling a trailer full of silage. He had been preceded by two fair-haired boys who had been helping him in the field and whose job was now to level the clamp. This far north there is only one silage cut a year.

Gunnar, a thin, pale man who rolled his own cigarettes and wore tinted lenses, looked more like a polytechnic lecturer than one of the most northerly farmers in the world. He took over from his father when he was 25 and has since built the herd up to 15 cows. 'There's no milk quota up here in the province of Finmark,' he explained. 'The government abolished it last year because milk production in the north of Norway has been falling.'

His Norway Red cows give 6,500 litres per year and, like all farmers in the extreme north of the country, he receives 60p per litre. The milk is collected twice a week and goes to the nearest dairy, 60 miles away in the town of Laksalv. 'You need a high price for your milk if you're going to survive this far north,' he explained. 'Don't forget that conditions up here get pretty rugged. Apart from the climate itself, we have nearly three months of darkness every winter and the cows can only stay out on grass in July and August.'

On a sunny July day I found it hard to imagine what winter would be like. How did he manage during the dark days? 'There are always lots of repairs to do, both to the buildings and the machinery,' Gunnar told me, 'and I have a computer which I use on the farm and as a hobby.'

What, I wondered, are the main problems about farming in the north? 'A long winter and a short hectic summer when I've got to do all the work you people would do in six months. The snag is that I've only got two months to do it in.'

There were other differences too. 'From time to time, usually in the early spring, farmers find reindeer grazing their precious grass. When this happens there can be problems with the Lapps who own the reindeer and who wander about the countryside with their animals. Some people have even gone as far as to shoot the trespassing animals, but this is a dangerous thing to do because there is a £3,000 fine if you're caught – not to mention what the Lapps might do!'

Had Gunnar ever considered moving south and having an easier life? 'Certainly not,' he replied. 'In fact I went to London this spring. I hated it. All that noise and dirt. And,' he added, 'Oslo's just as bad. No, I like it here.'

How did he see the future? 'Maybe I'll clear some more land by cutting down some trees and ploughing the bush so I can grow more grass. I don't really have enough land today which means I spend too much on concentrates. A few years ago the government would have paid me £4,000 per hectare to clear the land but they don't any more. Those were the days when they were encouraging us all to expand production.'

Even though Norway is outside the EEC and is untouched by the CAP, it became clear that the economic pressures which apply to me in comfy Cambridgeshire also affect Gunnar Olaussen on the edge of the arctic.

Ronny Mathisen was disconnecting the pto shaft on his forage harvester. The silage trailer was full and he was preparing to take it back to the yard to feed his cows who were waiting impatiently. It was sunny and the mosquitoes, which are ferocious in northern Norway, were already active even though it was only early afternoon. Mathisen's land was boggy and had already been cut up badly by his trailer tyres. The reason for the lack of drainage was a large river which I could see through the trees at the end of the silage field. But this was not just any old river; it was the Pasvik which marks the boundary between Norway and the Soviet Union. On the far bank striped red and green posts signified Russian territory, and watchtowers poked out of the forest at regular intervals. Ronny Mathisen is one of the few European farmers whose farm actually touches Russia.

Above the trees, less than a couple of miles away, I could see smoking chimneys in the Russian town of Nikel. The pollution put out by the nickel smelters had killed not only the forest on the hillside behind the town, but also thousands of hectares of the surrounding countryside. Compared to green Norway, Russia was a dead and sinister moonscape. It is lucky that the prevailing wind is westerly, otherwise Mathisen's farm would have suffered a similar fate.

Had he ever been across the river, I asked? 'Yes, five years ago. My wife and I went to a dance at Nikel just over there. It was a bit of a production as we were with a group and had to travel 80 kilometres through the official frontier post instead of just rowing one kilometre across the river.' Anyway, we invited them back again to a dance in Norway. I think they enjoyed it more than we did because there was a lot to drink.'

What did he think of his neighbouring town? 'Not much. It's filthy dirty. They hang newspapers in their windows instead of curtains. They are very poor.' But surely the vodka was good, I asked? 'No, not really,' was the unenthusiastic reply.

Was Ronny Mathisen nervous, farming only a few hundred yards from Russia? It was clearly a stupid question. 'Why should I be? I've farmed here all my life and my father before me. What is there to worry about?' I decided not to deliver a lecture about geopolitics. He was right.

Mathisen has 17.3 hectares on which he keeps 18 cows. 'Thank heavens they abolished the quota up here in Finmark last year,' he said. 'It enabled me to milk four more cows and has made quite a difference to my income.' His herd averages 6,000 litres per cow and, like Gunnar Olaussen, he receives 60p per litre.

Why, I asked, did he make both silage and hay? 'Silage is better because it is easier to handle, but I make hay for the calves and the sick animals. I also keep a bit of hay for my horse, Carino.'

A horse? At this point Mathisen disappeared, only to return a few minutes later mounted on a Palomino complete with western saddle. Wearing a red baseball cap instead of a Stetson, the Norwegian dairyman rode through his cattle just like a Texan ranch hand. I wondered what the soldiers in the Russian watchtowers were thinking as they watched us through their binoculars. They probably find Norway as confusing as I do.

Norway interested me not simply because it was outside the EEC, but also because the conditions (soil, landscape and climate) were so totally different to the UK. And I wanted to see the outer limits of agriculture.

THE WALK OF MY LIFE
1989

I went for a walk the other day. It was a walk I shall remember for the rest of my life.

The morning was cold, clear and crisp. The countryside was covered by a hoar frost which sparkled in the sunshine. Tramlines in the gently rolling fields of winter wheat stood out clearly and the oilseed rape was bushy-leaved. In the distance I could see a sugar beet harvester clearing up the last few rows.

The scene could have been anywhere in northern Europe, but there was one feature which made it very different indeed. Across the landscape, for mile after mile, ran an ugly line which took no notice of the terrain as it cut fields in half, crossed streams and slashed through forests. It was the Iron Curtain and I was in Germany.

The village of Hornburg was not damaged in the war and today its narrow mediaeval streets are still lined with timbered houses restored and painted as only a prosperous country like West Germany could possibly afford. Above the village, on an outcrop of rock, stands the schloss with its views in all directions over the steep tiled roofs of Hornburg.

Today the schloss is a farmhouse lived in by the Arnold family. Their daughter, Karolina, is studying agriculture at the nearby university of Gottingen. Her mother and uncle run the farm which consists of 525 hectares of arable land growing wheat, barley, rape and sugar beet. From the schloss's windows I looked over the fields and across the fence with its floodlights and minefields to where, less than two miles away, I could see the church spire of the village of Rhoden.

'I've never been there,' said Karolina, 'even though it's so close. The fence has been up ever since I was born, so only the older people in Hornburg can remember what Rhoden is like.' She paused to look at the familiar view. 'As a child I was always warned never to go near the fence because I might get shot. Our tractor drivers plough right up to the fence but they never speak to the guards on the other side, or even to the farmworkers from the State Farm who work the fields on the other side. 'I suppose,' she added, 'we've all got used to it.'

On the evening of 18 November, however, things began to change. The Hornburg Fire Department received a message which ordered them to go up to the fence and remove the gate which had blocked a small road leading east. The following morning the border at Hornburg was opened for pedestrians only.

The East German border has long been one of the most unpleasant and sinister places in the world. Silent soldiers with machine guns scowl as they search your car and confiscate magazines, books and cassettes of Mozart and the Beatles. On the return journey they look under the car with mirrors to make sure you are not carrying a fugitive.

I was therefore, more than a bit nervous when we set out down the narrow road towards the border. We passed a hut which once served as a shelter for the West German border guards. Taped to its window was a notice from the Red Cross advising visiting East Germans where in Hornburg they could have a cup of tea and change a baby's nappy.

The actual border is marked with a yellow, black and red striped post and there a few Hornburgers were chatting as they stared eastwards to where, 100 yards away, a group of khaki-clad East German soldiers stood in the shadow of a watchtower.

I tried to look calm and cheerful as we passed into no man's land and approached the border fence. Some bulldozer had smashed a hole in the wire mesh, and duckboards had been laid across the raked strip of earth which was designed to show the footprints of any escapers.

We approached the soldiers and Karolina, who must have been feeling quite as uneasy as I, wished them a good morning. They grinned and agreed that the weather was perfect. We showed them our passports and while they were examining the visas, I tried to pretend that this was just like any old border. After a few minutes we were waved through. I felt an overwhelming sense of relief before I realised that we had merely passed through an outpost. The proper examination was located half a mile away at the second fence, where a collection of portakabins and tents served as a makeshift frontier post.

Here, surrounded by all the horrible hardware of the fence, with its mine-fields, electrified barbed wire and searchlights, the soldiers were again cheerful and relaxed. They looked at my passport, stamped it and with some glee, announced that I was the first foreign visitor they had seen at their little crossing point. We were wished a good day, but as we strode off, I realised my palms were moist. Yet, we had done it. We had crossed the border into East Germany. Two weeks earlier, if we had somehow managed to reach the same point, we would have been shot. The thought made me feel slightly sick.

After a few hundred yards we stopped to ask two young soldiers the way to Rhoden. 'Three miles on the road,' they replied, 'but if you come with us we'll get you there in ten minutes.' To my amazement we set off up the concrete track which ran inside the second fence. This road, which had been built for military patrols, cut straight over the hill instead of snaking through the countryside.

As we reached the crest, a panorama opened up before us. To the left was Hornburg, in front of us the two fences with a wheat field between them, and to the right the village of Rhoden nestled in a shallow valley. I asked with great politeness if I could take photographs. The reply did not surprise me. 'No, this is a military area.' But then, after a quick conference, the elder of the two soldiers said, 'Oh all right then, I don't suppose it'll do any harm really.'

Encouraged by this reaction, I asked the soldiers if they had ever dreamed that one day they might be walking along the fence with an Englishman. 'No,' came the reply, 'it seems unbelievable to us.' It did to me too.

As we ambled down the hill towards Rhoden, I fell back a few paces. In front of me three young people were talking and laughing as they wandered through the countryside. Anywhere else in the world I wouldn't have given it a second thought, but gradually I realised I was watching history happen in front of me. A West German girl was laughing with two East German boys as they spoke the same language and shared the same jokes.

The first thing you notice about Rhoden is the smell. It is the smell of coal fires on a still winter day. The empty cobblestone streets are potholed, the houses need a coat of paint, the church – which had looked so impressive from Karolina's house – is deserted. By the bus stop a small poster announced the date of the football game. I cheered up when I noticed that the village's team is called Rhoden Tractor. It makes more sense than Sheffield Wednesday.

No traffic, no children, no noise. Only the sound of poultry showed that the buildings were occupied. In this ghostly atmosphere we ventured into what looked like a deserted farmyard. There in a corner an old man was washing two horses. Hermann Breustadt has worked on the same farm since 1936 and must have seen some dramatic changes. Today, of course, it is a State Farm and Hermann – long past retiring age – helps out with the 265 Holstein cows and is in charge of the six horses. That morning he had been trying to plough some private allotments but the ground was too hard and he had given up.

Outside in the sunlight four women had gathered round the village pump. When they learnt that we had come from Hornburg they began to get excited. Like the border guards, they too found it hard to believe that the border was now open. But, like the rest of the village, they had made the trip to Hornburg to see old friends for the first time in over 30 years.

It was time for a cup of coffee so we went in search of the village pub. This was not an easy task. Eventually, we found it attached to the village hall, or House of Culture as it is known in East Germany, but the sign on the door was not good news. Unlike any other catering establishment I have ever known, the Rhoden café closes on Tuesday and Wednesday. We were destined to remain thirsty.

In the garden of a shabby house an old couple were pruning the roses. Karolina, as usual, stopped for a chat. They too had been to Hornburg, where they had done some shopping and visited friends. They seemed almost bemused by their new freedom, as if they were dreaming and knew it could not possibly last.

At the edge of the village we stopped to look westwards across the two fences towards Hornburg. A few hundred yards away was Karolina's winter wheat and there in the distance stood the schloss with its tower. Karolina was silent as she absorbed a view she never thought she would see. I asked if she had a funny feeling in her stomach. 'No,' came the reply. 'Only that I am hungry. But, I do have a funny feeling in my knees.'

So did I.

I returned to Hornburg six months later and salvaged a piece of wire from the already decrepit fence. Today all traces of the border have long since disappeared. But I suspect that the inhabitants of Rhoden over the age of 20 must still dream at night about the bad old days.

EAST GERMANY
1990

I was looking at a pair of storks nesting on a chimney and didn't see the soldier. 'You can't go down there,' he barked. 'This is a border area and it is forbidden.' Horst Hundt, who was driving our little grey Trabant car, stiffened visibly. 'I couldn't give a damn what you say,' he announced. 'I'm a manager of the collective farm and these are our fields. Besides,' he added, 'I've got an English guest.' And with that he accelerated (if that is how to describe a Trabant picking up speed) past the soldier. There was a brief moment while we all pondered this encounter before Herr Hundt broke the silence. 'If I'd said that six months ago I'd have gone on a ten-year holiday. But today things are different.'

It was drizzling gently when we got out of the car and walked along the border fence to a deserted watchtower. Down a trench into a sunken entrance, through a heavy steel door which was already beginning to rust. Along a low passage past a primitive lavatory with a broken seat until, ducking our heads, we reached the foot of the tower. Clambering up a series of vertical ladders, we emerged at the top where we found ourselves surveying the north German plain with the Berlin–Hamburg autobahn in the distance.

In the centre of the roof the searchlight lever was still moveable. The only noise was the wind blowing through a cracked window. It was as if the last patrol had gone off watch a few minutes earlier. On a list of places in the world I had never expected to reach, an East German border watchtower, was near the top. As I looked west across the wire, the minefield and the patrol road, I realised that within a few years the towers, the fence and the Iron Curtain itself would disappear from the landscape and enter the history books.

In East Germany today you can feel history happening. It is not only in deserted watchtowers that you sense it. You can hear it when LPG (Collective Farm) directors talk about private property. You can see it on the tattered election posters stuck to every wall.

The headquarters of LPG Pampow is a long, low prefabricated hut. Outside a Russian army jeep was waiting while its driver had a slow cigarette. Inside, a secretary asked me to wait in her office; and a few minutes later a Russian officer emerged. What was a man with gold tanks on his epaulettes doing in a farm office? 'Oh him?' said the director, Dr. Helmut Nieter, surprised that I had even noticed, 'He lets me have some of his men to work on the farm and in return I provide his unit with vegetables.'

In the glass-fronted bookcase, flanked by potted plants and a visitor's book containing Julius Nyerere's signature, was a bust of Lenin and a copy of Eric Honecker's autobiography. Yet on his lapel Dr. Nieter wore the symbol of the recently formed Farmers Party. 'We got 7.5 per cent in Mecklenburg in the recent elections,' he told me, 'and I'm hoping to run for office myself in the local government elections.' I was moved by his passion for democracy, but later that day was told that in the recent past Dr. Nieter had had a reputation as a fearsomely left-wing member of the Socialist Unity (i.e. Communist) Party. Vicars of Bray are alive and well in East Germany today.

Coffee and sandwiches were brought by the secretary and Dr. Nieter began to explain – with some pride – the history of his 4,700 hectare farming empire which employs 200 people and runs 75 tractors. The soil at Pampow, as in most of Mecklenburg, is poor and sandy, suited to rye but little else. Like most collective farms, LPG Pampow was founded in 1953 after the Communist government had confiscated the large estates and 'encouraged' the peasants to put their land into a collective and become shareholders. Later, a directive from Berlin split all Collectives into separate arable and livestock operations. Yet today, 37 years later, the members (all of whom are invariably referred to as 'farmers') still nominally own the land and still receive an annual rent of £3 per hectare.

The original members of the Pampow LPG are long dead, and many of their children have left the land. 'Most of the workers,' explained Dr. Neiter, 'are no longer members of the collective. They are simply paid employees. We do, however, sometimes have a problem with land ownership. For example, the widow of an original member died this year and now there's a problem because nobody really knows who actually owns the land which the LPG farms. The original owner, the husband, never came back from the war. His widow lived on in the old farmhouse and his land went into the LPG. Today, the old lady's son has no proof that the land is his because it was never registered in his mother's name.' The future looks good for East German lawyers.

Out in the flat countryside with its thin sandy soil, I watched as a Russian tractor dropped off a set of rolls and went home for lunch while overhead a yellow plane was top dressing in the distance. LPG Pampow's biggest field is 203 hectares. I wondered when they had pulled out all the hedges.

'It's a shame really, they went too far,' replied Dr. Nieter. 'In fact we're now planting some hedges to reduce our fields to around 40 hectares. Also the hedges help as windbreaks, which is important on our blowing sand.' Wind was not the only problem at Pampow. Round a field of newly drilled peas they had put up an electric fence to deter wild boars.

Under the shade of a tree a tractor and cultivator were waiting for its driver who was eating lunch. Eventually he, and a man whom I took to be the co-driver, arrived on a bicycle. Siegfried Dornbusch is 55. His wife worked on a local estate before it was confiscated, and his own family used to own a small farm. I imagined he would be impatient to get his land back so he could enjoy free enterprise and the joys of peasanthood. I was in for a shock. 'Me, farm on my own?' asked Siegfried. 'Why should I?' This was hardly the reaction of a man who had laboured under the Communist yoke for his entire lifetime. 'Where would I get the money to buy the equipment?' he demanded. 'And besides, how could I manage today with 12 hectares. It isn't enough is it? You'd really need 50 hectares these days if you wanted to survive.'

I assumed I had stumbled on one of the few gormless members of the collective, but Siegfried continued. 'When the system changed last year and we realised that private property would be possible again, some of my mates were pretty keen on the idea, but in the past few months they've all changed their minds when they faced the practical problems.' What about the younger generation? 'Well, my son isn't interested, I know that. He's an electrician in Schwerin and he doesn't want to be a farmer. Mind you, he's keen to set up on his own as an electrician, but not on the land.'

Dr. Nieter's experience was not dissimilar. Over a sludge-like goulash in the farm's canteen, he explained, 'Only one man so far has asked for his land back, and he was a chap who wanted three hectares for his horse. I told him he'd have to take it all or nothing at all, so he decided to leave it in the LPG.' Dr. Nieter paused for another swig of coffee. 'In theory, at least, it is possible for people to come along and say that they want their land back again. If this does happen, it will effectively mean that the LPGs will be destroyed.'

Confident that his farmers would not rise up and demand their land back, Dr. Nieter was less certain that he could manage to survive the capitalist system. 'In the old days – until a few months ago – we got our orders from the State but today we're on our own,' he admitted. 'We don't really know what to do. It's all a little confusing because we are in a transitional stage.' I began to feel almost sympathetic for this man who had clearly been chosen to run a collective farm as much for his political reliability as his agricultural expertise 'What's worse,' continued Dr. Nieter, 'we don't get any money from the State. Instead we're expected to make it ourselves.' It was hard to conceal a smirk.

'We used to receive fixed prices for everything we delivered to the state and we were able to sell any surplus we had. Today everything is much less certain and life is going to be much less comfortable without the security the state offered. Take the lunch we're eating for example. It costs the workers one mark but it costs the LPG 4.50. Likewise the milk we sell fetches 1.71 marks per litre but the consumer only pays 0.69 per litre.'

LPG Grabow lies to the south of the Baltic port of Rostock. Its director is a 50 year old crew-cut man in leather jacket, who radiated energy and seemed quite at home talking about interest rates and profits. Horst Piechocki, who comes from Polish stock, was candid about the problems in the past. 'Sure I was a party member. It was necessary in my job. But we all knew the system wasn't working. We never had to think for ourselves, the central planners did it for us.'

'Take me, for example,' continued Piechocki, 'I've had three different jobs which the party ordered me to accept. I reckon I did well for the State. If I'd done it under capitalist conditions I'd today be a millionaire today. Of course, if I'd done badly I would have been sacked.' Was he a bitter man? 'Yes. Very bitter. I've given the state everything and look what I've got to show for it now.'

Unlike Dr. Nieter, Piechocki did give the impression that he had understood the fundamentals of capitalism. 'Our problem,' he told me quietly, 'is that we employ far too many people. If we are going to survive – and that is far from certain – we'll have to reduce the labour force by at least 50 per cent. Nearly half the people who work on the LPG are not concerned with agriculture. They do anything from looking after the kindergarten to building houses. The mechanics in the workshop – all 26 of them – are hoping to set up a business on their own. I hope they succeed. But the dilemma facing me is that I am elected by the workers, and they'll hardly vote for a man who is going to put a lot of them out of work.'

The sun was beginning to go down when we arrived at a building site. 'It's a bit of a fiasco,' admitted Piechocki, as we inspected a series of half-finished buildings. 'A few years ago the Ministry of Agriculture instructed us to increase our sheep flock and produce wool. Now we find that there isn't a market for wool and we've spent all this money building a complex for intensive sheep.' He grinned sheepishly and continued. 'With any luck we'll be able to rent these sheds to one of the new businesses which we are told will be setting up all over the country. We certainly won't ever use them for our flock.'

The rest of Grabow's livestock production had fewer problems. The Friesian herd was, I was assured, the third best in the country, with an average of 7,650 litres per cow. 'Our milk production has risen so much this year the local dairy has cut back our quota to 95 per cent of what we produced last year.' It all sounded a bit familiar. 'It should,' he continued, enable us to survive better when we get into the EEC.'

The pig unit seemed reasonable, and had figures which would not be amiss on Humberside. 'We shall certainly lose money with our pigs,' admitted Piechocki, 'because there simply aren't enough abattoirs in this country. As a result our pigs are butchered too late and too fat. Until we sort out this bottle-neck, things won't improve.'

'Starting on 1 May this year we will have no connection with the state at all. No quotas set by the government. We'll be on our own for the first time. Yes, I am a bit scared about the future. Its unlike anything we have ever known before and we are going to need a lot of help from the West.'

The field, like so many in East Germany, stretched to the horizon. On the headland a rather shaggy tractor driver was waiting to relieve his colleague on the sugar beet drill (who was driving too fast and very crookedly). I seized the opportunity to see if another potential peasant farmer was lurking under the beard. 'Certainly not,' muttered 31-year-old Lothar Tregar. 'Why should I want to farm on my own?' A shy man who clearly did not enjoy speaking to inquisitive foreigners, his real concern was the organisation of the farm itself. 'It's crazy,' he told me. 'I earn 600 marks a month basic, and yet other people with better and easier jobs than me earn a lot more.'

LPG Zarrantin, with 9,000 hectares, is one of the largest collective farms in East Germany. The reason for this was, however, political rather than agricultural. The western edge of the farm is the border with West Germany and used to be a highly sensitive area. This region was divided into four zones, the last of which was a strip five kilometres in which only specially selected people were allowed to live.

On foggy days the tractor drivers were forbidden to cultivate the fields near the fence lest they took advantage of the conditions to escape. The LPG's workers were not allowed to talk – or even wave – to the West German farmers on the other side of the fence.

A few years ago, after intense pressure from the West, the automatic shooting machines on the barbed wire were dismantled. Instead an additional fence, some 500 metres inside the original one, was constructed. The cost to the Honecker government was 1.4 million marks per kilometre which, by some strange coincidence, was exactly the price a kilometre of autobahn cost when Hitler built them before the war.

In the farm office I found Horst Hundt, who is in charge of agrochemicals and fertilisers. A 62-year-old man who, back in 1954 was a farmer with 34 hectares, Hundt is today a member of the LPG as well as being one of its top managers. Why, I wondered, had he given his land to the collective? 'In theory it was voluntary,' he told me, 'but in practice it would have been impossible for me to remain outside the LPG, so I suppose you can say that it was compulsory.'

Did Hundt want to return to being a private farmer? 'No. Not for me,' he said, shaking his head. 'I'm too old, but also I have no money and I would need a lot to start from scratch. Besides, 34 hectares wouldn't be big enough to support a family today. But most important of all, I don't know what the future will bring. You must understand that there's great uncertainty about everything. Life here in the GDR has changed very much in the past few months and who can say how it will change in the next six months? Most people are a little bit afraid still, and I am too. But only a little bit,' he added with a grin. 'So I'll stay where I am with the LPG. I've got three children. A doctor, a dentist and a mechanic. None of them want to be small farmers today.'

I wondered if any of the younger people on the collective had disagreed with him and asked for their own land back. Hundt thought for a while before answering gravely, 'I do not know of anybody.' Was he surprised by this? Another pause. 'No. Not really. We have a good life here and I think everyone knows it.'

Outside, Hundt's beaten up old Trabant with a Free Democrat sticker on the window seemed to sum up East Germany perfectly. As the rain continued to fall on an already gloomy landscape, Horst Hundt became philosophical. 'I've been with the LPG since the beginning of socialist agriculture. And now I'm seeing the end. I never thought it would happen in my lifetime.'

Today, Hundt is the object of a lot of attention on the LPG. The big western agrochemical companies are beginning to supply some farms in East Germany even though they are not getting paid in hard currency. 'They are giving us credit,' said Horst Hundt, 'in the hope that we will be able to pay for the stuff one day.' It reminded me of a drug dealer who gives away his wares, safe in the knowledge that once his recipients are hooked, they will come back again and again.

LPG Zarrantin grows all the normal arable crops, with more rye than any other cereal. It is split into three management units and employs a total of 270 people, of whom 180 are farmworkers. Zarrantin is an arable collective but provides forage for seven livestock collectives in the immediate area. I wondered, given this clear separation of responsibilities, what incentive there was to produce good silage and hay. The answer lay in the fact that all fodder is analysed for quality and the livestock collectives pay sliding scale.

We went outside into the drizzle, past rows of rusting machinery to where a potato store was being emptied. Rather than talk about spuds, I tried to find out if the hated Stasi (secret police) had been active in agriculture.

Hundt's answer surprised me. 'Oh yes, they were everywhere. Take this farm, for example. Out of the 270 members on the LPG, five or six will have been

Stasi informers. Today we don't know who they were and, personally, I'm glad we don't. But,' he added darkly, 'there are some people would very much like to know.'

Driving through the farm, we passed a series of centre pivot irrigators which reminded me of the American Midwest. 'They're Russian,' said Mr. Hundt. 'We have 19 of them and manage to irrigate 1,000 hectares of our sugar beet and potatoes. I wondered what the yields were on the irrigated land. 'Not all that special I'm afraid,' was the reply. 'Last year was the worst potato year we've ever had at around 19 tonnes per hectare and the beet managed only 30 tonnes.'

Horst Hundt is a frustrated man. 'Our problem,' he explained to me with infinite patience, 'is that supplies of chemicals and fertilisers are limited by the State. We are allowed exactly 120kg/ha of nitrogen for wheat and no more than 50kg of P and K.' He paused to make sure I had understood his fractured English. 'We've done trials ourselves, and we know we could increase our yields if we put on more nitrogen. The snag is that we also get a lot more diseases which we can't control because we simply don't have the fungicides.'

The Agrochemical Centre of Wittenburg serves four LPGs covering 27,000 hectares. Over a lunch of fishcakes and the inevitable gherkin salad, washed down by fizzy lemonade, the director, Christof Otto, was frank about the future. 'It's going to be difficult,' he told me, 'but at least we stand on two legs. On the one leg I am a contractor for the LPGs providing spreading and spraying machinery, and on the other leg I provide them with the fertiliser and agro-chemicals. We'll certainly have to be more flexible than we've been in the past.'

On the way round the plant I noticed a sign exhorting the workers to try harder in 'Socialist Competition'. Christof Otto shrugged his shoulders. 'Socialist competition only existed on paper, not in real life. Take us here, for example. Who were we competing with? We provide chemicals for LPGs and we have no competition at all. The regional authority decided how much each LPG should have and it was our job to distribute them. We were not allowed to make any decisions. Our only flexibility was deciding when and where to spray.'

The fizzy lemonade had loosened my tongue and gave me courage. Will the GDR farm managers be able to learn capitalism after a lifetime of being told what to do? Did they really understand things like interest rates, profit or incentives?

Horst Hundt, who had been silent for some time, burst into life. 'Some dogs,' said the 62-year-old man, 'remain young all their lives while other young dogs become old very quickly. The worst problem we have in farming in the GDR is that we employ too many people. About twice as many as necessary, I would guess.'

Next door to the Agrochemical Centre at Wittenburg was the Combine Repair Centre where 1,000 machines are overhauled each year. Inside, a line of Fortschritt combines was being hacked to bits to by an ant army of workers. My guide explained that the centre was forced to make most of their spare parts as they simply were not available from the factory. 'I suppose they have all been exported for hard currency,' was his cynical comment. On one workbench a pinup had been stuck to the wall. It was not a nude woman but a glossy picture of a John Deere combine.

LPG Scharbow was formed when, in 1979, five smaller LPGs (with names like Sunrise, Unity and Freedom) were amalgamated. Now the 6,500 hectares are run by three managers with independent machinery and labour. Over them sits the director, Mr. Nietrich, in a smart office with a fridge, a colour TV set, a sofa. On the bookshelf the inevitable mementoes of Russian delegations had not yet been removed.

I asked Mr. Nietrich what changes he'd like to see happen to his LPG under the new system. He had no doubts. 'I hope the livestock collectives will be given land of their own so that they could work independently,' he said fervently. As usual, there isn't a lot of love lost between the arable and dairy sides of any farm. 'Today,' continued Nietrich, 'we have to provide 700 tonnes of silage every day to the neighbouring livestock LPGs, and it isn't at all convenient.'

But it wasn't the problems of farm management which were worrying Mr. Nietrich, it was the future of land ownership itself. As usual, no LPG member had asked for his land to be returned, but the threat from west German farmers who had fled from their farms in the east was very real indeed. 'I don't know what will happen if these people claim that their land had been illegally taken from them,' said Mr. Nietrich. 'There will be a lot of political support for them in the west. I hear that the Bauernverband (the German equivalent of the NFU) is encouraging them to do just this and the politicians may be forced to agree.'

Was there, I wondered, a danger that foreigners will come and offer the farmers high rents – more than the £3 per hectare they receive from the LPG for their land? Mr. Nietrich pondered this question. 'Yes, this could happen, and when it does I suppose we shall have to match these rents. Of course, in the past this was forbidden in the name of socialist economics. Everyone had to receive the same pay, regardless of whether or not he had contributed land to the LPG.'

Outside in the yard a forklift truck was moving boxes of an unrecognisable vegetable. Only when I got close did I see that they were potatoes with chits at least six inches long. 'We've had a very warm winter,' explained Mr. Nietrich lamely, 'and it's given us a few problems with our seed potatoes.'

Inside a gloomy barn a riddling line was working under dim light bulbs. After my eyes had got used to the light, I noticed an extraordinary sight. Amongst a row of ladies wearing headscarves, a small group of Russian soldiers in full uniform and grey winter hats, were helping to pick over the spuds.

'You'll be surprised by this,' said Nietrich, 'because the West German papers say we hate Russian soldiers. It isn't true. They get on well here.' It turned out that the local Russian commander provided manpower in return for vegetables. Another platoon of soldiers was, I learned, out stonepicking that afternoon.

At this point I noticed that Mr. Otto had reappeared. He beckoned me over to a quiet spot. 'It's no good,' he whispered. 'The director is living in an imaginary world far removed from reality.' I wondered what he was talking about. 'Look at those seed potatoes,' he said. 'They're hopeless. As it happens, the LPG cannot sell half of them. Who'd want to buy these spuds with such huge chits? Would you? And yet Mr. Nietrich will tell you that he has sold all of them and his LPG is doing wonderfully well. It isn't. You should not believe everything you hear here.'

Otto went on to explain some of the contradictions of GDR agricultural economics. 'Take pigs,' he told me. 'Would you be surprised to learn that it is cheaper to feed them bread rather than barley? Bread is subsidised by the government so for 34 pfennigs you can buy 1.5 kilos of black bread. But if you want to buy a kilo of barley you'll have to pay 35 pfennigs. And it doesn't stop there either. I fatten eight pigs myself every year and I buy five tonnes of spuds. But I make sure I buy ware potatoes rather than stockfeed ones because they're cheaper. The housewife pays 20 marks for 100kg of potatoes while the farmer is meant to pay 24 marks for stockfeed potatoes. That's why I go to the shop and buy potatoes which were intended for human beings.' I was trying to absorb this extraordinary information when Otto continued, 'Some people in the GDR reckon that half the bread in the country is actually fed to pigs. That's what we call socialist economics.'

I was beginning to warm to Christof Otto, and felt a bit of encouragement was deserved. I told him that he would enjoy the EEC with its high farm prices and cosy subsidies. 'High farm prices?' exclaimed Otto. 'You must be joking.' I assured him that I wasn't, and began to explain how the EEC tariff barriers ensured that prices were well above world levels. 'Has nobody told you,' asked the incredulous Otto, 'that the price an East German farmer is paid is often double what you get in the EEC, and that as a result we shall have to accept a huge reduction in price, not to mention the fact that we are inefficient producers?' He went on to explain, as an example, that collective farms are paid over £200 per tonne for wheat.' I began to realise why all the LPG directors had a faintly nervous look to them as they pondered the future as part of a Greater Germany within the Common Market.

The Ministry of Agriculture is located in a leafy suburb of East Berlin. It occupies what was once a hospital and later a Russian Military headquarters. Today, like everything else in the GDR, it is gloomy, decrepit and dusty.

Dr. Wolfgang Beutner is Head of the Economics Department and had obviously risen to his important post by being a loyal servant of the old regime. Not surprisingly, his vocabulary and phrases contained much of the jargon so beloved of Communist governments. Fortunately we had the services of the moderately lovely and extremely talented Carola Appelmann, who has the rare distinction in the German Democratic Republic of being a regular reader of *Farmers Weekly* at the Institute of Agriculture where she works as a translator.

I confessed to being surprised that I had not yet met a single farmer who wanted to take his land out of an LPG and set up once again as an independent man. Dr. Beutner did not share my surprise. On the contrary. 'There was a public opinion poll taken recently,' he told me, 'which showed that 63.2 per cent of all LPG members wanted to stay inside the collectives, 4.2 per cent said they wanted to leave, and the remainder either didn't know or (more likely) didn't care.'

When the conversation turned to the future of GDR farming, Dr. Beutner had few illusions. 'If we went into the common market today there would not be a single LPG which would survive. They'd all go bankrupt because we simply don't have the technology which you do in the West. And, of course, the EEC prices are much lower than ours. This is why we'll need at least five years – and hopefully longer – to make the transition. Even then it will be difficult for our farmers.'

I suggested – with what I hoped was great diplomacy – that maybe, just maybe today's LPGs are too big. Ever the civil servant, he assured me that they were already being reduced in size and that the livestock and arable sides were – where possible – being amalgamated. In East Germany, he told me there is a proverb which says 'The boss's eye feeds the animals in the yard'. Quite how the Ministry of Agriculture squared this proverb with the 1,500 cow herds which occur on the LPGs is hard to understand. Perhaps the saying has only surfaced again with the departure of the Communist regime.

'Our problems are not simply the lack of technology,' continued Dr. Beutner. 'We employ far too many people and so our labour force will have to be reduced. I also believe that around 400,000 hectares, almost ten per cent of the farmland, is really too poor to cultivate.' This would, I reflected, be welcome news for setaside fans in Brussels.

The ancient but war-damaged city of Magdeburg sits at the centre of some of the finest land in northern Europe. The methodical Germans, who love to quantify everything, have decreed that the soil in the Magdeburger Borde rates a maximum 100 points.

The vast rolling landscape is dotted with piles of straw, constructed out of loose bales which have been shot by an agricultural version of a circus cannon into a heap. In summer lines of combines sweep across the plains, making Saxony look like Saskatchewan.

Just beyond the southern edge of the Magdeburg Borde, the village of Langenstein houses one of East Germany's leading plant breeding stations, where Dr. Albrecht Meinel welcomed me with the news that he had assembled no fewer than seven LPG directors to discuss farming with the Englishman. Over a table of chocolate eclairs and sticky cakes, we talked about the most profound political revolution to sweep Europe for 50, and maybe 150, years.

As managers, their perspective was inevitably rather different from that of a tractor driver. They were unanimous that they would be unable to survive unless they had large and immediate infusions of technology. But they agreed, albeit sadly, that it was unlikely that a Marshall Plan for GDR agriculture would assist them.

The group varied from those who were clearly products of the old system and who were beginning to realise that history had marooned them like whales on a hostile beach. Others, however, were evidently enjoying the prospect of freedom from the dead hand of the State. Yet even these men seemed a bit unnerved by the fact that no longer would the telex chatter and out would pour the latest instructions from Berlin. It must have been a comforting thought to know that the buck never stopped here.

Friedrich Kropf, who runs the Strobeck arable LPG, was very uneasy. 'Who is going to buy our produce? Indeed, will we be able to sell it? We don't have quotas, and the way things are going in the EEC why should they give us any when they already produce too much of everything?' Another of his colleagues chimed in. 'Our government is going to sacrifice agriculture for industry just as you've done in the West. Farming will end up being as useful as a fifth wheel on a car.'

Dr. Meinel confirmed what I was beginning to suspect. 'East German farmers today are like a rabbit looking at a snake. They are almost paralysed with fear and it means that they find it hard to think or plan rationally for a future which they simply cannot foresee.'

I suggested that some LPGs might start joint ventures with western farms, much as Volkswagen and Wartburg had already announced for the car industry. 'Oh yeah. It would work very well,' replied Joachim Luttge. 'It would be like the joint venture between the pig and the chicken. After they had agreed in principle they began to work out the details. 'I'll provide the eggs,' said the chicken, 'and you'll provide the bacon.' The pig thought for a moment and before saying, 'but that involves my being slaughtered.' 'Exactly,' said the chicken.

The meeting agreed one thing, however, which was that what GDR agriculture needs more than money is simply time. 'Give us until 1993 and we should be able to make the changes necessary,' said Peter Diedrich, who looks after forage at the Derenburg livestock collective. 'By then we should have been able to shed 50 per cent of our labour because many of the men are so old they will have retired by then. Mind you, we'll still have a few problems to solve even if we do have a smaller labour force.' In common with most of the livestock men round the table, he was well aware of the fact that huge herds and good stockmanship rarely coincide.

But what occupied their minds more than anything else – more than their lack of agrochemicals, more than their poor equipment, more than the evident overmanning – was the spectre that a West German landowner might appear tomorrow in his Mercedes demanding his father's land back.

'In my village,' explained Joachim Luttge, who runs a livestock collective farm, 'you must remember that many of the local farmers went west in the 1950s before the fence was erected. They are all alive and well in West Germany today and may well now feel that they can expect to be given their land back.' Remembering the deadened towns I had driven through that morning, it was hard to see why anyone would want to return.

I dined as the guest of the Langenstein livestock collective. This was clearly a successful operation which has made a lot of money, thanks largely to their turkey enterprise. They had invested surplus income – or what will henceforth be called profit – in a restaurant where, with a choice of warm lager or orangeade, we ate a Langenstein Farmer's Plate. To me it was a mixed grill.

That night I slept in the cathedral town of Halberstadt. The White Horse is an old merchant's house which is now a shabby and sad hotel. My room was a huge vaulted affair high up in the roof. I had a choice of four beds, one of which was a double, and a table which would happily seat ten for dinner. In addition I had two wardrobes, an industrial-sized black and white TV, a threadbare carpet and a single grimy towel. For this I paid the equivalent of £40. The price, I learned, had been increased by 400 per cent that month. East German hoteliers were learning fast about supply and demand.

Joachim Luttge is not a big man but is full of visible energy, whether he is talking or striding round the livestock collective which he runs at Derenburg, on the edge of the Harz Mountains.

'I started farming when I was 14,' he told me in his bright office with the regulation set of Erich Honecker volumes behind him. 'In those days my father ploughed with a horse. We had seven-and-a-half hectares with a few cows so I know what it is like to be an independent farmer. In fact my seven-and-a-half hectares are in the LPG today, so I am one of the few (10 per cent of the labour force) whose land is part of the farm.'

On our way to the farm I stopped to read a poster which was stuck to a lamp post. It advertised the German Democratic Farmers Party with the slogan 'Bread for the People. The town needs our farmers'. Dr. Meinel laughed. 'They managed to get 2 per cent of the votes in the GDR because they really only appealed to the old farmers and there aren't many of them left any more.'

We drove through the village to the dairy unit where 550 cows are milked as a single herd. They were tethered in two large sheds – cubicle sheds without the cubicles. The black and white animals were not Friesians as I had assumed, but were part Friesian-Holstein, part Jersey and part native Schwarzebund. The herd, which is zero-grazed, has an average of 4,500 litres per year. That morning they were being fed on fresh-cut forage rye. 'We're a month earlier than I have ever known before,' said Joachim Luttge, 'which has made our jobs easier than I had anticipated.' During the summer the herd is normally fed on a mixture of clover, grass and fodder rye, and in the winter on silage made from maize, grass and sugar beet tops. Dried lucerne and concentrates are distributed by hand, according to how much milk each animal was giving.

Outside in the yard Hans Schonebaum, one of the farm's five shepherds, was loading fresh-cut rye onto a small wagon which was pulled by what looked like a Shetland pony. 'The trailer belongs to the farm,' explained Joachim, 'but the horse, which is called Florette, is Hans's. He keeps it as a hobby and it's a lot cheaper than running a tractor.'

'The lambing this year was easier than ever, Hans Schonebaum told me. 'We must have averaged 150 per cent which is a bit better than our normal 135 per cent. The Mutton Merino sheep are raised primarily for their meat, but the wool clip is clearly better than we would get from mules in Britain. Out in the fields, I was amazed to see that many of the shepherds still dressed in the traditional Saxon costume with a hat, long black cape, sliver buttons and a thistle spud. At their side they had an Alsatian – which explains why these days they are often known as German Shepherds.

Inside the pig unit, where the German Landrace x Edelschwein sows looked healthy enough, Joachim felt the time had come to sound a muted note on his own trumpet. 'You know,' he began almost diffidently, 'our costs are only 75 per cent of our income which means that we made a 25 per cent profit. I'm very confident that we shall do well in the new economic situation which we face today.'

It was not the time or the place to point out that his costs were not real costs but were simply what the Ministry of Agriculture in Berlin had decided. He, like all GDR farmers, had been living in never-never land.

On the other side of the village a group of greenhouses marked the collective's horticultural venture. As we entered we were interrupted by a burst of machine gunfire. I flinched. 'Don't worry,' said Joachim, 'its only the Russians. They have a training ground nearby.'

Tomato plants were growing well, having just replaced lettuces in the beds. In due course they would be followed by ornamental flowers. Outside gangs of women were potting small tomato plants by hand. Joachim admitted the system was antiquated but they did not have the cash to buy the automated machinery.

'We built all these glasshouses with our own labour,' said Joachim with justifiable pride, 'and we only started in 1986. The trouble is that we have to heat the glass with brown coal which is not very efficient even though we use automatic boilers. But any day now we shall be able to change to oil. That will be much more efficient. But first we need money.'

'But first we need money.' These words kept repeating themselves to me as I drove east, past Colditz Castle to Dresden and the Polish frontier. LPG directors, who were chosen more for their political reliability than for management skills, now assume that all they need is money. They have a childlike belief that, given western machinery and lots of chemicals, they will survive, and even flourish, in the EEC.

Their naiveté is as touching as it is misplaced. Many LPGs will surely go bankrupt before they reappear in some very different form. Banks will be reluctant to lend money and foreign investors would be well-advised to hold off until the whole agricultural system settles down in a few years. In the meanwhile a new breed of manager will take over the farms and maybe – just maybe – a man will ask for his grandfather's land back.

I had come to East Germany full of certainties, and was leaving full of doubts. I remembered the great square of Halberstadt, with its Gothic cathedral at one end and a stupendous Romanesque church at the other. On a weekday evening at 7pm it was completely deserted. I shut my eyes and tried to visualise what it would look like in five years time. The shiny new parking meters would all be full and the pavement cafés would be bursting with people and pop music. Sandwich stalls, sausage stands and shops stuffed with souvenirs would complete the picture. Which of the two visions was preferable?

Albrecht Meinel's optimism was well placed and he now runs a successful plant-breeding station for a West German company. Most of the other characters in this tragedy were less lucky. The cooperatives were sold off and their land was usually – but not always – let to hyper-efficient West German farmers who arrived with John Deere tractors and Claas combines. Their first act was to make all the workers redundant, which explains why rural unemployment in eastern Germany today is still over 25 per cent. It also explains why the villages I visited are today very sad places.

HUNGARY
1990

The sour cherry soup was cold and refreshing. The conversation was stimulating and the atmosphere more cultural than agricultural. 'Beethoven visited this house many times,' remarked Dr. Laszlo Balla, referring to the castle in which his office is located. I took another sip of the dry white wine from Lake Balaton and reflected that this was not what I had expected to find in Eastern Europe. I was in Hungary or, to be precise, at the epicentre of Hungarian arable farming, the Research Institute at Martonvasar.

Dr. Balla, in a black shirt and a white tie, looked more like a trendy vicar than Hungary's leading wheat breeder. Over lunch he told me about his background. 'I came from a peasant family in the north where we farmed ten hectares. Like all farms at that time, it was entirely self-sufficient. As well as growing wheat, maize, potatoes and sugar beet, we also kept three cows, two horses, a handful of pigs and a few chickens.'

I asked whether the farm was today part of a cooperative. 'Yes,' replied Dr. Balla, 'but not because my father ever agreed. In spite of all the pressure which was put on him by the Communist authorities, he persistently refused. Eventually he could hold out no longer. He gave up farming in disgust and took up his other profession of carpentry. My grandmother was the one who took the decision to give up the land, the livestock and the equipment.'

As we talked about the bad old days of the postwar Communist regime, Dr. Balla became pensive. He gazed out of the window into the elegant park which surrounds the house. 'In spite of my family's experiences, you shouldn't overlook the fact that a lot of Hungarian peasants were pleased when the collectives were started in 1947. They weren't at all unhappy to give up their land and join.'

He took another sip of wine. 'The fact is that life in those days was neither easy nor pleasant. For example, my father's ten hectares consisted of 35 separate strips of land, each of which was far too small for a tractor. In my village, which had a population of 2,000, there were only two bathrooms. One was owned by the priest and the other by a prosperous engineer.' There was another pause while he remembered some distant detail of a forgotten childhood, before returning to the present. 'Today, my mother still lives in the same house and she has two bathrooms, gas central heating and a colour television.'

Another sip of wine. 'And the improvements weren't just social either. From an economic point of view the collectivisation of Hungarian agriculture had some great advantages. It enabled us to use machinery instead of horses or oxen. I know it has not worked well, but you must not forget the good aspects of a bad system.'

Suitably chastened, I finished the excellent sausages and contemplated a plate of fresh strawberries which had been set in front of me. Coffee and Hungarian brandy ended the meal. I was going to enjoy Hungary – always assuming I managed to stay sober.

Dr. Tamas Bodizs has pale brown eyes which match the colour of his leather jacket. They also stare out of his thin face with a ferocity which I found slightly unnerving. As president of the cooperative farm at Szekesfehervar, he was entitled to both an attractive secretary and a large office. Like his Czechoslovakian counterparts, it too had a padded leather door, but the walls were bare and there was no sign of the certificates showing the successes of socialist agriculture, or the souvenirs deposited by visiting Russian delegations.

Over a glass of Schweppes sparkling orange, I sat back while a torrent of facts, opinions, jokes and more facts poured forth. Three thousand five hundred hectares of wheat, maize, sunflowers and lucerne, 1200 head of cattle, including 300 dairy cows, 200 sows, 300,000 broiler chickens and 40 million eggs produced each year. In the middle of this catalogue the windows rattled as a series of explosions shook the air. 'The Russians are practising again,' he remarked with studied nonchalance. 'They have a training ground next to the farm and it sometimes gets a bit noisy. It's also a bit inconvenient when they are holding manoeuvres because they shut the farm roads and we almost have to steal our own crops.' There were, however, a few compensations. Whenever one of the farm's tractors got stuck, a passing tank could always be hailed to pull the machine out of the bog. The normal currency for such a transaction was a bottle of Palinka, the fierce Hungarian brandy.

Had Tamas seen any dramatic changes as a result of the new politics in Hungary? 'No. Not really. Remember, that we've been moving in this direction for the past ten years, unlike the other Central European – not Eastern European please – countries. The only real difference is that as a result of opening our borders, there are now export possibilities there never were in the past, when we could only sell to the East Bloc countries.'

This development was not, as I had imagined, an unalloyed blessing. 'We must now be very careful,' he told me, 'because Hungary is full of men from Western Europe who try and buy our produce and sell us things without any real financial backing. They are fraudulent people looking to make quick money at our expense.' I informed him that we call these characters cowboys. Tamas pondered this information for a while before announcing gravely, 'That is a very good name.'

But it was not just the wide boys from the west who were worrying Tamas. 'We have now pulled down the Iron Curtain and the western countries, particularly the EEC, have told us that they will support us. At first I was grateful, but then I realised we were being fooled. You people from the EEC do nothing of the sort. You erect tariff barriers which make it impossible for us to sell any of our agricultural products. Take our eggs, for example. When I try to sell them in the EEC they bear a tax which raises their price by 50 per cent. How am I supposed to compete in that sort of world? I get a price of around £50 a tonne for my wheat, but when I buy farm machinery or chemicals I have to pay the prevailing world market price. How can I possibly survive like this?'

Like most passionate people, Tamas took a few minutes to get warmed up. By now he was in his stride. 'The other day we had a delegation of high-powered French politicians and they asked us what sort of help we needed. They thought they were in the Third World and were offering us a tractor or two. I told them that we could do without their so-called aid. All we needed was a fair chance to trade on equal terms. I don't think they liked to hear that.'

I wondered how he felt about Hungary's neighbour, Austria, which is not a member of the Common Market. The reply was immediate and emphatic. 'Austria? It's worse than the EEC. They protect their farmers even more than Brussels.'

At last the tirade was over and the Schweppes fizzy orange (a welcome change from the indigenous pop) had all been drunk. It was time to go out into the fields. In the farmyard, where a lot of Claas and John Deere machinery contrasted with the normal East German Fortschritts, a grey Polish Fiat was waiting. It was equipped with a CB radio into which the President kept issuing instructions as he drove me at enormous speed over rutted farm tracks. A fractured skull seemed inevitable, even with the primitive seat belt which the Poles thoughtfully provided. Past hedgerows full of courting couples (the dust cloud we left must have made them extremely uncomfortable) and fields of sunflowers drilled in straight rows to the horizon, we bumped and thumped and crashed.

Had anyone on the farm asked for his land back? 'No, not yet,' shouted Tamas over the noise of an imperfect silencer. 'In fact we had a meeting the other week and it was agreed unanimously that everyone would leave their land in the cooperative. But,' he paused with a grin, 'they all said that they wanted more money.'

A field of lucerne was being harvested by a self-propelled Hesston. The driver, Janos Kovacs, was waiting in the sunshine until a second lorry arrived from the grass drier to run alongside and take another load. Did he own any land? 'I suppose I do,' he admitted after some thought. 'My mother had seven hectares. It was given to the farm many years ago but now I could get it back.' Did he want to? This time his reply was immediate. 'No. Not at all. Where would I get the money to put up a building or buy the tractor?'

Had Mr. Kovacs actually thought about the possibility of becoming a farmer himself? 'No. Not really. There isn't a lot to think about really. Besides, it's much too early. I'll have to see how things turn out in the next year or so before I even begin to think those things.' By this time the lorry had run alongside and it was clear that I was holding up the harvest. We disappeared in our normal cloud of dust, with me crouched down low, fearful of another spine-jarring smash.

Tamas Bodizs smiled a private smile and I asked what he found amusing. 'The real reason Kovacs doesn't want to get the land back is because it isn't his decision. His mother is still alive and we pay her money each year for her land. It's like another pension for her. But somehow he didn't tell you that.'

Suddenly the car stopped and Tamas leapt out. 'There!' he cried, flinging his arms wide. 'A 300-hectare field full of lovely wheat. Are you in the West really saying that it would make sense to split it all up into two hectare lots. Who would be the winner?' This rhetorical question had barely been posed before we accelerated off across the level Hungarian landscape.

In the distance I saw a small cloud of dust which turned out to be a flock of sheep, accompanied by two donkeys, two goats, a small black dog and an old man. Ferencs Petu is 66. 'I was a private shepherd with my own flock for over 20 years,' he told me, 'but when I retired I couldn't bear to be without animals, so I kept 100 ewes and now look after them. They keep me company.'

Why the donkeys? 'To carry hay.'

Why two goats? 'When I was younger and had a flock of 500 sheep I used to graze military airfields far from civilisation. I kept two goats so I could have some milk each day, and now I hate the idea of a flock without goats.' While we talked, the small dog (called Gypsy) herded the donkeys as well as the sheep.

It was milking time for the farm's 250 Friesian-Holsteins, who were filing through a 26 standing polygon parlour. 'The dairy operation makes money,' remarked Tamas, 'which is more than I can say for the beef unit. The prices are so bad that we really run it as a charity to keep jobs for some members of the cooperative. Don't forget that a member has a right to a job for life, so in a way it's quite useful having beef cattle, even though they show a loss.'

In the nearby village we stopped at a neat house surrounded by a garden containing a plastic tunnel. 'This is where one of our employees lives,' explained Tamas. I met Istvan Czeiffra, who is foreman in the cooperative's pig unit. Together with his ten-year-old son, Balasz, he was feeding the eight sows which he keeps in the back garden. The boy dragged me over to look at a Duroc boar which he then proceeded to ride like a Shetland pony round the small yard. 'He's called Cork because he looks like a bottle cork.' And so he did.

It was time to go home, but before we did so, Tamas had one more place to show me. We parked the car outside a door sunk into a small mound on which a bungalow was being built. 'One of our workers has started a new enterprise,' he told me proudly. 'Come inside.' Opening the door, I peered into a black void. 'Mushrooms,' said Tamas, and he switched on the light to reveal rows of plastic bags growing button champignons. I admired them politely and turned to leave the tunnel. 'I must tell you the truth,' said the president coyly. 'They are mine. And that is my house which is being built above us. Three years ago I wouldn't have dared tell anyone that a cooperative farm president had another activity because it would have been considered bad form. But today everyone thinks it's an excellent idea. I shall try to start a marketing cooperative with other growers and we shall sell the mushrooms in Austria. But that is a few years away. In the meanwhile I shall go on learning the business.' It was a mark of how Hungarian farming is changing.

The Hungarian Parliament building in Budapest dominates the Danube even more than the Houses of Parliament dominate the Thames. The Ministry of Agriculture looms over a noisy square nearby. Like so many Eastern Bloc ministries, it is decorated with a taste which may have appealed to Stalin, but left me feeling queasy. A statue of a noble but rugged peasant, with a water bottle in his hand and a scythe on his shoulder stares out across the traffic. The ten-foot figure seemed very different from Mr. Kovacs on the forage harvester yesterday.

Entering through the large wooden doors, I found myself in a gloomy lobby which looked like one of the smaller London railway stations. I had an appointment with Dr. Janos Mezei, deputy head of the department. But to reach his office on the third floor I had to take what the Hungarians call a 'Pater Noster'. This is a primitive form of lift consisting of two continuous lines of boxes. It means that you have to step nimbly into the small box as it moves past. This was great fun; I spent five minutes going round and round while civil servants got on and off at the different floors.

Mr. Mezei, cutting a dashing figure for a civil servant, in a dark turquoise blue jacket, was politeness personified. We began by talking about the problem of land ownership. There was, I learned, a new bill going through Parliament which would become law in August. Called the Land Act, it sets out the framework under which private farmers can take their land out of the cooperative farms and set up on their own. The theory of returning land to its previous owners may be simple, but the practical problems are mind-boggling.

'You must always bear in mind,' said Mr. Mezei, 'that only land which was taken by the State after 1947 will qualify to be returned. This means, in effect, that only farms of 100 hectares or less will come into this category since all the big estates had been confiscated before 1947.' He also pointed out that since the average size of Hungarian farms at that time was a mere six hectares, few of today's owners would find themselves with enough property to make a viable farm.

A tribunal had been appointed to try and sort out the inevitable problems which will arise when people realise that they will not actually receive the precise piece of land they had owned before. They will, however, receive an equivalent area of the same quality, thus ensuring that a large field of 100 hectares is not split up to provide the six hectares which once made up a peasant holding.

Forty per cent of the land in Hungary is owned by individuals, although it is today farmed by cooperatives. A further 40 per cent is in the name of the cooperatives because the original owners have either died or moved away from the farm. Former owners of this category of land – providing they can prove their title – will also be eligible to have it returned, or will be able to sell it to the cooperative at a price to be fixed by the tribunal.

Among the many headaches is the fact that since 1947, 800,000 hectares of farm land, much of which is still nominally owned by individuals, have been lost to roads and building. The owners will receive equivalent areas of land and will not suddenly find that they are the owner of a factory, shopping centre or part of a motorway.

And yet in spite of the problems of land reform, Mr. Mezei was optimistic about the future. 'In ten years I predict that 80 per cent of the land in Hungary will be farmed by private farmers,' he announced with a faintly triumphant smile. His definition of private farmers did, however, need a bit of interpretation, since he assumed that most would be working in small groups within the struc-ture of today's huge cooperatives. Thus 20 men might group together to milk 250 cows through an existing parlour, using existing buildings and machinery. They would own their own animals and hire the facilities.

An alternative system would be for today's cooperatives to turn themselves into limited companies, with the members receiving shares which they could buy or sell.

When we turned to the economic outlook, Mr. Mezei was less optimistic. 'I realise all too well that Hungarian agriculture is inefficient because we employ too many men. But what can we do? The cooperatives have a legal duty to provide their members with employment for life and we can't alter this legisla-tion. We've inherited this from our socialist past and we can't avoid it. At least today our labour costs are low compared to yours, but what will happen if wages rise to Western European levels is anyone's guess.'

I stepped out into the sunlight from the gloomy ministry, and hurried off to a suburb of Budapest where the Crop Protection and Conservation Service is based. In a small office I met Attila the Hungarian, hunched over a slide projector and bursting to tell me how they classified soil mineral deficiencies, and other exciting organisational problems. A large, shambling, cuddly man of 47 with a droopy moustache, Attila Fekete could not have pillaged a village, let alone an entire nation. I was looking at an anti-climax.

After a good lunch of goulash, Attila and I piled into a minibus and drove out into the country where one of Hungary's most famous farms is located in the village of Herceghalom. The HKG State farm is probably best known for its Hungahib hybrid pigs which have won prizes across Europe. The 4,000 hectare farm also milks 1,000 dairy cows and manufactures agricultural sprayers.

Gabor Gondocs reminded me of a middle-rank Velcourt manager. Bright, energetic, arrogant and unstoppably talkative. He had the answer to all problems and was afflicted by neither doubt nor uncertainty. And, more than anything else in the world, he wanted to make a good impression.

Having classified me as an Englishman, he knew just what would interest me. 'We have,' he bubbled, 'a joint venture with RDS in your country for spraying control mechanisms.' When, later that afternoon, I was taken to the sprayer assembly line, I found small groups of men cutting up bits of metal outside in the farmyard. It was all most picturesque but hardly what I had expected. 'We are having an open day next week,' explained Gondocs, and the building where we make the sprayers has been emptied so we can use it as a conference hall.' He went on to say that they were expecting 20 foreign companies to exhibit their equipment and 'for the first time we shall be showing small sprayers suitable for the independent farmer.'

The dairy operation was as impressive as the sprayer factory was quaint. One thousand cows are milked three times a day through two herringbone parlours side by side. 'We copied the technology from California,' said Gondocs. They seem to have copied the milk yields too, if the herd average of 7,200 litres can be believed.

It was time to say goodbye to Attila the Hungarian. His handshake was probably as firm as that of his ancestor but his shy smile was a great deal friendlier.

It is hardly Szeged's fault that it is not a pretty town. In 1879 the river Tisza burst its banks and left only 300 buildings standing. In a small and dusty sidestreet I was welcomed by Ferenc and Imre Dobay. We went through a gate into what appeared to be the back garden of a suburban house. It was as if I had been magically transported from a grimy Hungarian town to Boskoop in Holland. I found myself standing in a vast greenhouse full of geraniums, fuchsias and begonias.

'My grandfather founded the business, which specialised in selling carnations,' explained Ferenc, 'and my father continued until 1952. At that point our land was confiscated by the Communist government. We were considered to be Kulaks, or rich peasants, and life became very difficult. In fact I found it hard to go to the local school as I had been branded as an anti-social element.'

Three years ago, when the liberalisation of the Hungarian economy was well underway, Ferenc decided to try to start up the family business once again, with the help of his brother. He has clearly succeeded. 'It hasn't been easy because the government was not at all sympathetic to private undertakings, but we were lucky here in Szeged – the city council didn't ask too many questions.'

It is lucky for the Dobay brothers that cash sales comprise 90 per cent of the turnover. The remainder is with state organisations who pay by cheque. 'We have been very careful to make honest tax returns,' said Ferenc, 'and as a result we have been left alone. But there are some people in this town who aren't so honest. Today they own three Mercedes cars.'

Life had not been easy for the family. 'You probably don't understand,' said Ferenc, 'because in Britain a horticulturist is a horticulturist. But here in Hungary you cannot just go out and buy a greenhouse, or seed trays or any of the other things you need to set up in business. We have had to build everything ourselves. We have had to find the glass, the wood, the pots, the compost – and pay high prices for all of them because the state was not in favour of this sort of activity.'

Today, under a government which encourages private enterprise, the Dobay brothers are expanding aggressively. Imre took me to his house which, complete with a two-car garage, had been completely rebuilt. In the garden three large plastic tunnels, heated by gas, were full of geranium cuttings, and on the ground floor he showed me, with justifiable pride, a laboratory he had recently completed to produce geranium clones. 'All the equipment was bought second hand,' he pointed out when he saw me looking at a microscope.

His optimism was infectious. 'The other day we had a visit from a man who owns a chain of garden centres in Vienna,' he told me. 'He is going to buy plants from us because they are better and cheaper than the ones he used to buy in Austria.' The fact that the Hungarian border had recently been opened made this transaction possible whereas a year earlier it would have involved selling the plants through the state-owned monopoly.

Over a cup of strong black coffee, the brothers admitted that they were now beginning to think about getting the old family holding back from the State. 'It isn't exactly straightforward,' said Imre, 'because although the old house still remains, much of the land has since been built on and is now a housing estate.'

And yet in spite of the hardships which the Dobay family had endured since 1952, they were still amazingly optimistic about the future. 'We horticulturists don't need subsidies or special grants from the government,' said Imre with characteristic Hungarian passion. 'All we want is to be allowed to function as a private business and not have any artificial restrictions placed on us. Given this sort of freedom we'll do well.' I didn't doubt it for a second.

Szeged is famous for its peppers, which provide the basis for Hungary's most famous ingredient, paprika. That night I dined on a splendidly hot fish soup, a fiery paprikash (a solid yet delicate sort of goulash) and ended with the local schnapps from a bottle containing, inevitably, hot peppers. The after-taste felt as if my throat had caught fire. Throughout the meal gypsy violinists played to tables packed with singing Hungarians. It was an evening to remember.

My liking for Szeged was increased both by the number of heavily laden cherry trees and by the notice in my hotel room which read, 'If you would like to phone, just pick up the receiver and wait till the operator enters. You can tell her all your further wishes.' The use of the word 'all' was, I reflected, typically and generously, Hungarian.

Ruszke is a small, dusty village on the southern edge of the great Hungarian plain. In the potholed main street I saw more horses than tractors, and the head-quarters of the local cooperative farm was equally unprepossessing. A telex machine chattered away in the corner of a dark office where Imre Palvolgyi met me. He was keen to practise his scanty English. 'My wife is leaving for your country next week. She will be living in Shropshire with the Young Farmers.' I assured him it would be a memorable experience and that she would be completely safe among the young farming gentlemen of Shropshire.

He seemed to be relieved, and began to describe the farm on which he works as an agronomist. 'We have 3,000 hectares, but one third of this is forest. We grow the usual crops like wheat, maize and sunflowers, and some you may not have in England, like paprika and garlic.' It turned out that the cooperative used to grow no fewer than 350 hectares of paprika, but this has now shrunk to a mere 15 hectares.

I wondered what had caused so sudden a decline. 'Quality,' replied Imre. 'We have now realised that paprika is a crop which should really be grown by private farmers because they look after the fruit so much better.' This was slightly sur-prising coming from a cooperative farm manager. 'We found that when we picked the paprika by machine, most of the peppers were damaged and the quality was very low. Nobody wanted to buy them and so the price was also very low. If a small farmer has half a hectare of paprika his family pick them by hand and take great care of everything. The quality is good and the price is high. That is why we now have concentrated on growing corn for seed and have almost stopped growing paprika.' It was the first example I had come across in Eastern Europe of the market place playing a role in agriculture.

I declined the offer of a trip to the milking parlour since Imre had already confessed that it was rather primitive. 'We are not a rich cooperative farm and have not had any money to invest in our cows.' Instead we went out into the flat countryside where the watchtowers of the Yugoslavian border guards showed that we were as far south as it is possible to be in Hungary.

The huge wheat fields were showing signs of drought, with brown patches spreading outwards from the thinner soil. 'The drought has been very bad this year,' said Imre, 'and we are now expecting a wheat yield of around four tonnes per hectare, which is not at all good.'

In the distance a yellow helicopter was spraying fungicide, helped by a tractor which moved along the headland to provide a marker for the pilot. 'You must see our irrigator,' said Imre. 'It is from America and we like it very much because it covers 250 hectares in a single circle.' A few minutes later the largest centre pivot irrigator I have ever seen hove into sight. At its centre a group of men lounged around a sprayer and tried to look busy when we approached.

On the edge of every village small houses, each surrounded by a plot of land, showed that the inhabitants did a bit of farming on the side. 'In the past year we have had 40 requests from a labour force of 300 to have some land returned to them,' said Imre. 'The amounts are very small, averaging half a hectare each, which means they are only required for part-time farming.' Most of the cooperative members in this part of Hungary are effectively part-time farmers, working for the cooperative during the day and on their plots in the evenings. They borrow the cooperative's machinery to sow and harvest their grain, and buy straw from the cooperative to bed their animals.

Would the cooperative survive in its present form? 'No. I don't think so,' replied Imre. 'Most people are waiting to see what the new laws about land ownership will say and then they will set up on their own. In this part of Hungary, because the climate is good and the soil is fertile, they won't need more than ten hectares to live fairly comfortably. Of course they will have to grow mainly vegetables rather than wheat. The cooperative will change radically and simply provide facilities for the local farmers, who will all be independent within a couple of years. I think,' said Imre pensively, 'that our cooperatives will soon be like the ones in Holland.'

Dr. Borthaiser was in a hurry. 'I can only see you for half an hour,' he told me, 'because there are many things to do.' With six hectares under glass to supervise, I believed him. On his desk a computer terminal flickered away. 'It is from Holland and it controls the temperature in all the greenhouses,' said the bespectacled doctor. He led me out to where, on one side of a central passage, roses stretched into the distance, and on the other ladies were harvesting gerbera, which looked like small dahlias.

The greenhouses are heated by geothermal water which is pumped up from deep wells at 80 degrees Celsius. I assumed this was a cheap source of heating but Dr. Borthaiser corrected me. 'This unit may be cheap to run, but it was very expensive to build. We have five wells and each is over a mile deep. Imagine what they cost to drill.'

I was surprised to see two large greenhouses full of green paprikas, hanging from bushes which rose to nearly six feet high. Why, I wondered, did the cooperative grow this vegetable under glass when it seemed to flourish so well outdoors?

'Hungarians want to eat paprika throughout the year,' explained the doctor with infinite patience to a foreigner who clearly did not understand the Magyar soul. 'We can produce them in February here but they are only ready to pick in July if they are grown outside.' It was obvious really, but I had failed to understand that Hungarians are obsessed by paprika. I know now.

Benak is 36 but looks older, which isn't really surprising when you realise that he works 16 hours a day seven days a week. The reason for this schedule is that he is that rarest of Eastern European animals, a private farmer.

When I arrived at the farmstead, Benak's sister, Elizabeth, was harnessing the mare to a cart. Her job was complicated by the fact that she was also attached to a young foal which had been born this spring and would not leave the mother's side.

I assumed that a horse and cart meant that the farm was primitive, but I was wrong. 'We have three tractors,' said Benak, which I've bought cheap from the local cooperatives when they were about to throw them on the scrapheap. I've fixed them up and they work well.' He also had two balers, a forage harvester and a large trailer, which seemed a mite excessive for a herd of 28 cows.

Benak had spent four years working on a cooperative farm when, in 1982, he decided to set up on his own as an independent farmer. What made the decision even more odd was the fact that he did not own any land himself. He managed to rent a total of 20 hectares, some of which are water meadows beside the local river.

How had the authorities reacted to this decision to set up as a capitalist in a communist society? 'It wasn't easy,' replied Benak, 'but it was not actually illegal. They tried to put pressure on me and made life difficult by continually inspecting the farm to see if I was breaking any laws. But I wasn't and so I just kept going. Then they kept changing the regulations every year – and sometimes every month – just to make things difficult for me.'

'In those days I only had a building for the cows and my own two hands. Nothing else. Every forint I earned I put back into the farm so I had almost nothing to live on myself. That's where all the machinery came from.'

What had his mates on the cooperative farm thought? 'They reckoned I was crazy, but at least they were nice enough not to tell me this until it was clear I was going to succeed, some four years later.'

At this point an old man, wearing a beret and three days' growth of beard, stumbled out of the house. Benak's father is 72 and helps with odd jobs around the farmyard. Did he approve of his son's adoption of free enterprise? The old man muttered a few words and relapsed into silence. 'He doesn't say much,' remarked Benak, 'but he doesn't like what I am doing because when he was young he saw all the land taken away from farmers like me and he is worried that it might happen again.'

It was time for milking. The old man went into the cattle yard with a whip to which he had attached the red, white and green Hungarian colours. After a lot of shouting and whip-cracking, he persuaded the cows to come into the barn. I walked through the farmhouse, which was quite exceptionally squalid inside, to the milking parlour which occupied a space on the end of the building. Inside I found I was in one of the best and cleanest parlours I have ever seen. An Alfa Laval bulk tank and six standings from half a herringbone were clean and modern, in total contrast to the rest of the farm and the house.

Hungary is full of oddities.

Another was when Dr. Peter Erdei, Director of the Szeged Cereal Research Institute, was talking about how the new regime had changed life in Hungary. 'The economics have not altered much, because we started our reforms ten years ago. But now we do not see the parachutists any more.' Parachutists, it turned out, were the Communist appointees who were foisted on organisations and drifted down from above. It was an expression I shall remember.

More than half of Hungary is covered by the Great Plain, or Puszta. It was once the home of wandering herds of cattle, who were tended by native cowboys on horseback. Today the Puszta looks more like Kansas than central Europe, with fields of wheat, sugar beet and sunflowers stretching to the horizon. But occasionally I caught glimpses of what the landscape must have looked like before the arable revolution struck Hungary. I also saw why this country is second only to France in the production of foie gras. Goose farms, large and small, flourish in Hungary.

Long, low, white barns with thatched roofs still stand out on the plain, far from any village. Isolated cottages, with their tall crane-like wells are still occupied. On the dusty streets of dusty towns, where horses outnumber tractors, men drink palinka (rough alcohol) and women wear headscarves. You still see sights which disappeared two generations ago in England, like a herd of pigs escaping down the main street late in the evening; the whole village turned out to drive them back.

At the northern edge of the plain, where the flat land gives way to upland valleys full of wild flowers, stands a town whose name is known throughout the world wherever wine is drunk. Tokay, on the banks of the River Tisza, produces only white wine. These range from the famous sweet Aszu to some rather heavy dry wines which reminded me of unfortified sherry. The villages in the area are evidently prosperous, with neat houses and the occasional stork nesting on a high chimney. The small private cellars built, like air-raid shelters in the gardens, may have fallen into disrepair, but the State Farm maintains a massive cellar consisting of a network of small tunnels dug into the hillside.

It was in such a labyrinth that the cellar master showed me thousands of barrels slowly maturing in the damp and winding passages. I had sampled 15 different wines and was feeling benevolent and a little bemused. I stumbled over a small step which was hidden in the shadows and remembered a conversation in Budapest a few days earlier. Dr. Janos Mezei was describing the confusion facing agriculture after the Communist regime. 'Hungary,' he said, 'is in a dark tunnel somewhere between the past and the future.'

Of all the eastern European countries, Hungary always had the most efficient agriculture. Under what was sometimes called 'goulash Communism' the farms were allowed tobuy western technology, which explains why today so many British farmers have invested in Hungary. The wines of Tokay have improved massively, thanks to French expertise. The one place which has probably got a lot worse (and a lot more expensive) in the past ten years is that restaurant by the Danube at Szeged. I still dream about the hot fish soup.

POLAND
1990

Jan Kwiatkowski is desperate. Beneath the dovecote in his tidy farmyard near the eastern city of Lublin, he explained the problem which obsesses every Polish farmer today. 'A year ago a litre of milk bought me 2.5 kilos of concentrate. Today it buys half a kilo.' The new Polish government, in a brave (maybe foolhardy) attempt to replace the old command economy with a free market, has dragged the country from Communism to Thatcherism overnight. In doing so, it has triggered off the sort of inflation that would make a Brazilian wince.

Inflation in Poland is so high it is calculated monthly. In January 1990 it was running at 40 per cent per month and the government hope that by midsummer it will have fallen to 7 per cent. The evidence is visible far beyond the Polish borders. Cars, loaded to the gunwales with toilet paper, toothpaste and tuna fish, head east along the German autobahns.

But inflation is not the only problem facing Polish agriculture. Unlike East Germany, where the farms are too big, Poland has the opposite problem. Averaging less than five hectares each, peasants' holdings comprise 80 per cent of the land. What is worse, one in five farmers is past retiring age. Poor Poland.

Poor Jan Kwiatkowski. He, like all Polish farmers, is suffering more than the rest of society, whose salaries keep roughly in step with inflation. 'Six months ago,' he told me,' I was milking 11 cows. Today I have seven, and if things continue as they are, I won't have a single cow by Christmas.' Yet, in spite of seeming so unhappy, he is strangely optimistic. 'It is the price we must pay for freedom,' he told me. 'I am sure that things will get better. The old system was no good.'

Jan Kwiatkowski's troubles aside, there are some compensations for a wandering foreigner like me. Petrol costs 90p per gallon and an on-the-spot speeding fine (assuming the Lada police cars ever catch you) is 25p. And then, there are the Poles themselves. Like the Irish they are Catholic, charming, cheerful, argumentative, sad, romantic and feckless. Only one thing worried me: the faintest distant echoes of anti-Semitism still seem to survive in a few dark corners.

The town of Wroclaw will not win any beauty contests today – which is hardly surprising since it was almost completely destroyed in the war. For 700 years until 1945 it was a German city called Breslau and was the capital of Silesia. Today, the farm buildings in the flat countryside around Wroclaw still look decidedly Germanic.

In one of these farms, consisting of a white house with a well in the farmyard and a collection of slightly dilapidated buildings, I found Josef Delikowski and his wife feeding their chickens. Two turkeys and a goose, which had just hatched 22 goslings, completed the poultry department.

The farm itself amounts to seven hectares, half of which is down to cereals and the rest to oilseed rape and pasture. 'I used to grow sugar beet but stopped a few years ago,' Josef told me. 'There was too much hand labour involved.' Today the farm is smaller than it used to be. 'We have just one cow left,' Mrs. Delikowski remarked sadly, 'And I think we'll sell her this summer. She's nine years old and manages to give only ten litres a day. But she should be worth a million zlotys for meat and the cash'll come in handy.' The same fate was not, however, going to befall Cubar, the 14-year-old horse which the Delikowskis keep in addition to their tractor. 'She's as much a pet as a worker,' explained Josef, 'but she uses less fuel than the tractor and is better at light work.'

Yet in spite of their complaints, the Delikowskis own their land free from debt, have a tractor, a car, a comfortable house and a colour television set. But these possessions had not come cheaply. Had they, I wondered, ever taken a holiday? 'No. And what's more, I haven't even considered it since we were married in 1956,' said Mr. Delikowski. His wife was, as usual, even more emphatic. 'The only holiday I've ever had was when I went to hospital a few years ago.'

'I can't wait until I retire,' continued Mr. Delikowski. 'We'll give the land to the state in return for a pension, but we'll keep the house and one third of a hectare. As far as I'm concerned, retirement can't come soon enough.' His wife nodded.

Jan Dereswiszcki is 42, but his grey hair and lined face make him look a lot older. Until 1984 he was a factory worker, but when his father-in-law died, he inherited five hectares and decided to become a farmer. Since then he has bought an additional ten hectares from the state without borrowing a single zloty from the bank and, as a result, is one of the biggest farmers in the area. But far from being optimistic and expansionist as I had expected, he was deeply unhappy. 'It's been a disaster,' he told me. 'I'd sell up and go back to being a factory worker like a shot if anyone was stupid enough to buy this farm from me – but of course they won't. And who can blame them?'

After tying up his particularly vicious Alsatian, Jan led me into a dilapidated shack. Like most peasant cowsheds, it was dark and my eyes took some time to pick out three black and white cows (he milked seven last year) which shared the accommodation with five sows. 'I'm going to sell all these animals,' he told me angrily. 'The way things are today there's no point in keeping any. I've poured all my money into this place and done it all myself. I've restored the buildings, put in a mechanical milking system, bought two tractors and a combine, and what do I have today? Nothing.'

He swung open a barn door to reveal a new and very large Polish-made Bison combine. How, I wondered, could he justify such a machine on a farm this size? 'I'm a contractor as well as a farmer. I suppose I cut 100 hectares last harvest and I'll do the same this year. The trouble is that with this inflation I daren't tell anyone in advance what I'm going to charge.'

After this catalogue of disaster, I expected a good old whinge about the politicians from Jan Dereswiszcki, but I was wrong. 'I don't blame today's government so much as the old party hacks who are still in the administrative jobs they've held for ages,' he said. 'They're the ones who have made this mess.' It was a complaint I was to hear often over the next week. A large part of the old Communist core, or nomenklatura as it is known, is still in position simply because there is nobody else qualified to do the job.

Drive south from Wroclaw and you soon begin to climb into the Sudeten mountains which divide Poland from Czechoslovakia. It was in this apparently primitive part of the world that the first sugar beet factory in Europe was started by a German back in the 19th century.

After driving down avenues untouched since before the motor car, past superb country houses left to decay, I found myself in a peaceful valley. The village of Jaszkowa Dolna is as pretty as Wroclaw is ugly. Dominated by a church with an onion dome, it is bisected by a small stream running down the main street. On either side clusters of sturdy farm buildings show that the land is fertile and the farmers were once prosperous. Outside each farm a single churn waits to be collected.

It was in one of these farms that I found Eugeniusz Babiak trying to start his tractor. The normal method had apparently failed and the machine was standing silent at the foot of a steep slope. Ever since the starter motor stopped working three years ago, Eugeniusz always ensured that he parked the tractor at the top of the hill so it would roll down and bump-start.

He was standing beside the tractor in the middle of the most squalid farmyard I had seen outside Africa (though I was due to see worse during the rest of my stay in Poland). A few diseased chickens pecked around on the dungheap and a turkey guarded what I assumed was the workshop. In the corner a well showed every sign of still being used, but seeing (and smelling) what occupied the rest of the farmyard, I would not like to have sampled the water.

Behind me was, once again, the remains of what had once been a prosperous German farmhouse. Eugeniuscz, sporting what is called designer stubble on Oxford Street, was politeness personified. He explained that his parents came from the eastern part of Poland which was occupied by Russia after the war. When the Polish borders moved 120 kilometres westward, his family, along with thousands like them, were given land from which the Germans had been evicted.

An old man emerged blinking into the sunlight. 'Talk to him,' said Eugeniusz, 'he was one of the first farmers here after the war. He'll tell you what it was like.' The old man shuffled from foot to foot. 'It wasn't easy to start with,' he said. 'In fact this part of Poland was often called the Wild West because it was so dangerous. There was nobody living here so bands of robbers made the law themselves, and in some of the forests gangs of German soldiers still survived. It wasn't fun, but in some ways it was easier to farm then. I remember we got 100 zlotys for a quintal of corn and we paid our workers 20 zlotys.' Eugeniuscz, who had been growing increasingly restless, could contain himself no longer. 'Don't be stupid,' he said, 'your memory's playing tricks with you. Besides, that was all 40 years ago.' And so it was.

A 29-year-old bachelor, Eugeniuscz farms 11 hectares. His arable crops consist of wheat, which he sells to the local co-op, and mixed spring corn which he feeds to his cattle. As we talked about rotations, he suddenly grew animated. 'This year I'll grow sugar beet for the first time,' he told me proudly. 'I've grown fodder beet in the past but never sugar beet.' He proposed planting half a hectare and would hire a contractor to drill the crop. All the other work, including harvesting, he would do himself by hand. Was there, I wondered, a quota for sugar beet and milk in Poland? 'Quota?' The word was foreign in every sense.

'No,' he replied after the concept had been explained, 'we can grow as much as we like and produce as much milk as we can.'

The dairy operation consisted of six cows, two heifers and six calves. Fifty litres of milk was collected each day by the local dairy. I have long been used to farmers boasting about their wheat yields, lambing percentage or milk production, and I had noticed a solitary ten-litre churn sitting on the bench outside the farmyard. Might Eugeniuscz not have been exaggerating his herd's performance? 'Not at all,' he replied. 'It will soon be Easter and I'm keeping back most of my milk to make cheese which I'll sell privately for the holiday.'

By now Eugeniuscz was convinced that I was not a government spy and began to relax. As I examined the barn for signs of its former occupants, he ran into the house and returned with two faded photographs which dated from the 1920s when the farm was owned by a German. The only evident relic of the past was a crucifix in the garden with a German inscription. It was clear that no repairs had been made in the past 50 years, and the buildings today were on the point of terminal collapse.

Nova Ruda, like all coalmining towns, is extremely ugly. But around it lies some lovely upland countryside. Among the small valleys is the hamlet of Scinawka Srednia, dominated by the rotting remains of a great schloss which once housed the German landowner. Today it is part of a State Farm and, like many of the best houses in Poland, will probably not survive the century unless a lot of money is spent on restoration. On the far side of the valley one farm stood out. Not because it was higher, but because of the unfinished new house which showed there was money in the family.

Jan Pajdziak was not a well man. He slumped in the dark farmhouse kitchen looking morosely at his grandson who was playing on the floor. Aged 67, Jan had recently retired, and was now suffering from bad lungs and, to make things worse, a broken arm. His youngest daughter bustled around making me a cup of coffee which looked like brown Windsor soup and tasted worse. 'I bought the place in 1962 from the State,' explained Jan in a soft whisper, 'but we've split it because my daughter here married a coalminer and wanted some land to build a house on.'

Jan's son, Woijtek, was evidently an energetic chap. He had built the new house behind the farmyard with a splendid view across the valley to the schloss. I was beginning to be impressed with a seven-hectare farmer who could afford a house like this, but there was an explanation. Like so many Poles, Woijtek had been to America and, after working as a mechanic in Chicago for two years, had saved up enough dollars to pay for his new home.

I found him squatting by a clamp throwing small potatoes onto a trailer and large ones into a bucket. 'We'll sell the big ones to a hospital and use the others as seed,' he explained. 'I usually buy my seed in from outside, but last year we had the best potato harvest we've ever had and I thought I'd try for it again with the same spuds.' The yield, I later discovered, amounted to 45 tonnes from a single hectare.

Looking up at the surrounding hills, I wondered how high we were. 'I don't know,' said Woijtek, 'but we must be over 350 metres because we get the special hill farmer livestock subsidy. We've got five cows, eight bullocks, nine ewes and fifteen lambs.' He had not included the fair-sized flock of chickens which were swarming over, through and under a large but tidy dungheap in the yard.

Why, I wondered, did a man who could have remained a mechanic in the United States, want to become a farmer in the mountains of Poland? 'That's a good question,' replied Woijtek. 'I ask it myself very often but don't have an answer. And,' he continued, 'when you realise what inflation's doing to us, you won't understand either. Take that tractor (he pointed to a small Ursus). Last year it cost three million zloty and this year it costs fifty million.' So how was he going to survive? A shrug of the shoulders was the only reply, but it managed to say it all.

The road continued to climb steeply and, though it was a warm spring day, snow still lay on the northern slopes. Eventually I emerged onto a small plateau and saw a cluster of wooden houses in a clearing. The hamlet of Pasterka was once a proper village; today there are only seven houses, most of which are occupied by widows. Bumps in the green pasture show where the farmsteads used to be, but the remaining buildings, although rickety, were well built when the German farmers constructed them 75 years ago. On some houses the original wooden gutters were still working.

Stanislavska Jaworska was exactly how I had imagined a Polish peasant to look. Small and tough, her brown leather face included a beak nose and a smile which lit up the hillsides. A headscarf, leather jerkin and boots showed that she was still working at the age of 75. She put down her pitchfork and beckoned me into the dark, low-ceilinged kitchen. The room was dominated by a large tiled construction which, on closer inspection, tuned out to be a combination kitchen range and central heating boiler. The temperature was tropical and the noise too was faintly reminiscent of the jungle. The explanation was that the old lady was rearing 50 chicks in a pen next to the heater. 'I'm waiting for the weather to get warmer before putting them outside,' she explained. 'I'm worried about buzzards. They eat baby chicks, you know.'

We began to talk about her history. 'My late husband and I were born in central Poland and came here after the war when the Germans went, so I've been here most of my adult life. In those days there were 70 farmers here in Pasterka, but many of them drifted away, frightened that the Germans were going to return.' Stanislawska paused to pick up a chick from the melée at her feet. 'Many of the farmsteads were deserted, which meant that the land had no owners. But then the State Farm took it over and they now use these old pastures as summer grazing for their animals which spend the winter in the valley.'

Had farming changed since those early days in Pasterka? Stanislawska stared out of the window as she tried to remember her youth. Then she pulled herself together with a start. 'We have never owned a tractor in our whole lives. Never. In those days we farmed 13 hectares and grew wheat, barley and fodder beet for the animals.' Wheat? I was surprised because we were over 1,000 metres high and I wondered how it was possible. 'Possible?' asked Stanislawa in her firm voice, 'Possible? There were no problems. We did everything by hand in those days but later we had horses. Actually, we did have an ox when we got here so I can't pretend we did everything by ourselves.' And she thumped the table with a gnarled fist as she enjoyed her little joke.

I wondered how they managed to harvest the wheat. 'Well, the ox pulled a reaper and we'd go along afterwards and tie up the sheaves.' Yet even in those days, I learned, they had electricity and mains water here on the top of the mountain. The Germans looked after their farmers well and the Poles are benefiting to this day. 'It is funny,' said Stanislawska, 'but a lady from West Germany has sent me food parcels on three occasions and has asked me to keep the buildings in good condition. I think she must have been born here.' I wondered, somewhat pessimistically, whether she might not also be anticipating a return to what many Germans still consider to be rightfully theirs.

Today Stanislawska has retired and lives with her son, a forester. The farm consists of one hectare of pasture which two cows, one calf and a bullock graze. Two horses provide the motive power. From a small pen a lonely Merino sheep stared out at me. 'I keep her for wool,' Stanislawska told me. She went into the next room and returned bearing two sweaters. 'I have knitted these from the sheep's wool,' she said to me with some pride.

Were there any wild animals in the forests? 'You probably have a lot of wild boar in England, don't you?' she replied. 'Here they are almost tame; they come right up to the house looking for potatoes. We also have very many badgers, some deer and all the other normal things you have in forests.' Any wolves? 'No. And no bears either.' I was very disappointed.

In central Poland there are more horses than tractors, which is not entirely surprising since this is the poorest agricultural area in the whole country. The carts are invariably low-slung four wheel, rubber-tyred models which, depending on the load, need one or two horses. I found them somewhat nerve-wracking because the carter sits so low he cannot see over his horse. Thus they tend to turn across the oncoming traffic, oblivious to approaching cars. The horses all have red bobbles attached to their bridles (and sometimes to the whips) because there is an old Polish superstition that a red ribbon keeps a horse healthy.

As I moved east, I noticed that field work became more labour intensive. I watched while four people walked down a furrow planting potatoes by hand while a horse plough kept a few yards ahead of them.

In the main square of Kielce a flourishing private market was in full swing. When it started earlier this year the police tried to prevent it but now they have given up the struggle. Yet many of the merchants didn't like having their photograph taken. One van was selling eggs. I assumed they were from private farms but was assured by the salesman that they were, in fact, the products of a State Farm. If you can't beat 'em, join 'em. This particular State Farm was learning fast.

The Rural Solidarity office occupies what used to be the Communist Party newspaper building on the edge of the square. In the local MP's office glossy pictures of western farm machinery covered the walls – souvenirs of his recent trip to England. My guide was a bearded young man who was keen to practise his English. 'I shall be coming to England this summer,' he chirped. 'I shall work on a farm near Canterbury which makes me happy because I have studied *The*

Canterbury Tales by Geoff Chaucer.' This was, I felt, carrying intimacy too far, and braced myself to hear about Bill Shakespeare and Charlie Dickens.

The casual motorist in Poland today does not often see a real village. They are usually by-passed by the main road, leaving the old settlement on either side of an unpaved street. I turned off the asphalt road and bumped down the village high street trying to avoid the potholes. As I did so, a hawk swooped down and picked up a young chicken before flying away to enjoy the meal.

Jan Zieliniski was waiting for me at the track which led to his modern farmhouse. He took me into what appeared to be the basement. 'Be careful,' he said. 'She kicks a bit.' I was standing next to a mare with a very small foal. 'It's less than a week old,' said Jan proudly. At this point his wife appeared, followed by his seven-year-old son, whose blonde curls and bewitching smile beggared the description 'angelic'.

The Zelinski farmhouse was built over the animals' quarters. Sharing the accommodation with the horse were three sows and a total of 28 piglets. 'They are American,' said Mrs. Zieliniski, pointing to some brown piglets which clearly had Duroc blood.

Jan is a stalwart of the local Rural Solidarity. He farms 12 hectares, milks four cows, and, like most Polish peasants, lives simply but in some comfort. 'My in-laws built the house back in 1967 and put the animals in the bottom because it was cheaper that way. But,' he continued, pointing to a new building in the farmyard, 'I built the barn all by myself.' I noticed that another site had been levelled and wondered what he was going to erect. 'I had hoped to have a cowshed there,' he told me, 'but we don't have the money and it will have to wait. It is one of the problems of inflation. We cannot spend money because we don't have any.' Why not borrow it from the bank? Jan grimaced. 'It would be the same if I bought some rope to hang myself.'

I had heard a lot about so-called self-help cooperatives in Poland and wondered whether Jan's village had organised one. 'We've thought about it,' was the answer, 'but it takes a lot of money to set up. One of the local farmers has a lorry and he now buys agrochemicals direct from the producers instead of buying from the State outlets. In this way he saves money and sells them to us cheaper.' The first faint signs of entrepreneurial flair were already evident in the Polish countryside.

Jan had not finished. 'Today we are inheriting the problems of communism,' he continued. 'Nobody has been trained to buy and sell in this country for 40 years so it isn't surprising that we are not very good at it today. We need help so we can do it like you in the west. Then things will get better for all of us.'

We walked back across the farmyard in which chickens, geese and turkeys all co-existed in something like harmony. It was time for the moment I had learned to dread – the Ceremony Of The Meal. Polish farmers are exceptionally hospitable, and the only way they can show this is by offering you a large meal. This takes place regardless of time of day, and irrespective of whether you have already eaten that day. On this occasion I was facing my third meal since midday.

We retired to the dining room where a table had been set for the male members of the party. The ladies gathered in the kitchen and peeked round the door to ensure that I was eating enough. Tomato soup with vermicelli came first, followed by pork, cabbage and delicious fried potato. This was washed down with glasses of strawberry juice containing whole strawberries. Eventually a dish of chocolate pastries appeared and we were expected to munch our way through these before ending with a glass of coffee or – in my case – tea.

Barely able to get up from the table, I found myself being asked a question I thought I would never hear when visiting a foreign country. 'Would you,' enquired Jan, 'like to see a very bad farm?' Having spent weeks looking at pretty good farms, I readily agreed, and we set out across the hillside and down to the village.

As we approached, I realised that we were heading for what can only be described as a hovel. 'The farmer is very ill,' explained Jan, 'and he has been victimised by the old Communist authorities, so it is not his fault that his farm is bad.' This was a substantial understatement. The cottage looked as though it had not been occupied for a decade. Gutters were broken, tiles were smashed and the windows had been patched. In the garden a heap of straw surrounding an apple tree caught my attention. 'Look,' said Jan. 'Pigs live here.' And sure enough, beneath the tangled and rotten surface I could see a few pigs. They were, at least, ideally placed to eat the windfalls. In the corner of the garden a real fairy story well, complete with rope and bucket, completed the scene of rural squalor.

By this time the farmer had stumbled out of the house and fetched his horse for me to admire. Jan greeted him in the matey way that local politicians know so well, and explained that I was a visitor from England. He also fished from his pocket an official-looking form which he gave the old man. 'It is from the doctor,' he explained, 'and means that he does not have to do any work.'

We left Jan for our next appointment in a nearby village. As we picked our way through the potholes, three men swayed towards us. They were carrying bottles and were very obviously in the last stages of inebriation. 'That,' said my guide, 'is a Polish pub. We do not have pubs like you do in Britain. Instead people drink anywhere. And sometimes they drink too much.' It was impossible to disagree. Throughout rural Poland prostrate figures asleep in the hedgerows show the power of vodka.

Josef Jajasnik was a bit impatient. A sallow 31 year old, he owns 11 hectares and rents two from the state. As well as being a farmer, he is also the local chairman of Rural Solidarity. His branch was holding a meeting in the village hall that evening and it looked as if Josef would be late. Because of this, I was profoundly relieved to learn that there was no time to have yet another large meal, so instead we ate home-made brawn and sausages with enough garlic to knock them dead in Warsaw.

In the gathering gloom, we walked down the village street past one of Josef's neighbours who had a young roe deer in the garden, among the pigs and poultry. 'He caught it by accident last year,' explained Josef, 'and now it is a pet.'

The meeting was attended by 14 of the local party stalwarts in a neon-lit room like many English village halls. 'Our problem,' said Josef, 'is what to do about the former Communists who are now trying to tell people that they are members of Solidarity. We have local elections next month and these people would like to be elected so they can go on doing the same jobs that they have done for years. We must make sure that they are not, but this is not as easy as it sounds.' Voices were raised as everyone expressed their opinions in a typically noisy Polish manner.

It was a peaceful spring afternoon without a trace of a wind. In front of me stretched water meadows which lined the bank of a small river. In the distance I could see a man and wife top-dressing wheat by horse. The man steered the usual Polish four-wheeled cart while the wife sat on the back broadcasting urea.

It looked like any other view in eastern Poland, but this was special. The river was the river Bug, which marks the border between Poland and Russia. On my side I was surprised to see that the Polish frontier post was empty. Not a single soldier was there to tell me to stop. I drove up past the red and white barrier to where a metal bridge crossed the river 50 yards wide. A watchtower made me remember East Germany. On the Russian side the fields were enormous, without a sign of life. On the Polish side I counted five tractors and three horses all working in a mosaic of small fields. It was a good example of the difference between the two systems of agriculture.

The villages which line the River Bug were poorer than any I had seen in Poland, but ornithologically they were richer. On many electricity poles storks had nested and were hatching their eggs. The presence of a river and water meadows meant frogs – the staple food for storks.

This is one of the few parts of Poland where land lies derelict and unused. The reason is partly economic (the farms are small and the younger generation does not want to work four hectares) and partly social (problems between Polish- and Russian-speaking populations). The soil is thin and sandy, growing rye, potatoes and clover, with the rye yields rarely exceeding one tonne per hectare.

It could not be true. I must have imagined it. As I sped down yet another of Poland's avenues, I thought I had spotted a machine I had never seen before. I turned the car round and drove slowly back to the cottage where a young man was cleaning out a very small drill. I had been right. Outside the front door, beneath an apple tree, stood a reaper. Not a binder but a reaper. In the western world they exist only in museums.

'It's very old,' said Henrik Madig. 'My father bought it in 1950.' I assumed he meant 1850 and asked my guide to repeat the question. 'No,' said Henrik with some exasperation, '1950. We use it every harvest.' The farm consisted of five hectares, two cows, one horse and a tractor. The horse was used in preference to the tractor because it cost less to run. 'I only drive the tractor,' explained Henrik 'when I am ploughing. For all other jobs the horse is better.'

Henrik was the youngest farmer I had met in Poland. Would he, I wondered, like to have a bigger farm with more land? His answer surprised me. 'No, I am happy the way I am today. It would make for problems if we had more land.' How about more animals? 'Yes, that would be good but they would cost a lot of money and we cannot afford them.' No nonsense in Poland about easy credit.

Later that afternoon, when the conversation flagged, my guide asked if I knew the chairman of the 'English Farmers Union'. I admitted a nodding acquaintance with Sir Simon Gourlay and was slightly surprised that the name did not elicit a flash of recognition. 'No,' was the reply, 'he is not the chairman of the English Farmers Union.' There was a pause while he searched in his wallet before triumphantly producing a visiting card. 'See,' he said. 'This is the President of the English Farmers Union.' The name on the card was Barry Leathwood, National Secretary of the Transport and General Workers Union. I decided I would not try to explain the difference between the T and G and the NFU. Only in Poland could Mr. Leathwood be confused with Sir Simon.

Marian Vnuk was standing by his Massey Ferguson (built under licence in Poland). Behind him was the old wooden cottage in which his 86-year-old father still lives, and the unfinished two-storey house which he has been building since 1978. Marian's welcoming smile sparkled in the sunlight because most of his teeth were golden. In honour of our arrival he had shaved well. Too well – his chin was still bleeding. Besides being a 10-hectare farmer, he has two other claims to distinction; he owns two tractors and no horse.

'I hope you do not mind that we have no meat today,' he apologised as we prepared for The Ceremony Of The Meal, 'but it is Good Friday and, of course, we are Catholics.' Over a gigantic plate of scrambled eggs and doughnuts, the conversation turned inevitably to politics. 'It wasn't easy in the old days,' Marian told me. 'Back in 1981 when Solidarity was founded, we had a lot of problems from the police. In fact, I led a strike which occupied the government buildings in Lublin for nine days. It was only ended when the government introduced martial law. After that I went to prison for 48 hours.' It all seemed like ancient history to me, but to Marian it is still part of his life.

'In those days,' explained Marian, 'the farmers who were members of the Communist Party had an easier life. They could get spare parts for their machines and building materials for their houses. The reason my house isn't completed yet is that I was known as a Solidarity man and so could never get bricks or timber. Today I can get the materials but there is a problem: I just don't have the cash.'

Jan Kwiatkowski is also a member of Farmers' Solidarity. In the tidy farmyard his 17-year-old son, Peter, was polishing his motorbike with a friend, while above him doves flapped in and out of the dovecote. 'There's no profit in milk any more,' he told me. 'It is even worse than arable farming, and that's pretty terrible too.' I was clearly speaking to one of the Lublin district's leading farmers. His herd average of 5,160 litres would not disgrace any western European dairyman, and the standard of his farm buildings was better than I had seen anywhere in Poland. 'I'm waiting to hear if I can export my calves to Italy or Greece for veal. If this is possible, I'll sell them at two weeks old which makes a lot of sense in today's inflation.' He paused to swat a fly which had settled on his cheek.

'The tragedy is,' he continued, 'you now can buy anything you want in Poland – providing you have the cash. A year ago it was exactly the opposite. No matter how much money you had, there was nothing available. Unless,' he added with a grin, 'you happened to be a member of the nomenklatura (the inner circle of the Communist Party). For them life was easy. It's not the same now. They're all living very quietly hoping that the old regime will return.'

Did this mean, I asked, that he no longer supported the Solidarity-dominated government? 'Oh no, not at all,' replied Jan. 'It is necessary what they are doing, but I'm frightened that they do not understand what life is like in the countryside any more. They are cutting down the tree of agriculture which took so long to grow.' I could hear many farmers in the EEC agreeing with this statement.

The Ministry of Agriculture is housed behind a huge classical facade on Wspolny Street in the middle of Warsaw. I arrived there on Easter Saturday when the churches were full (indeed there were queues outside) and the offices empty. It took me ten minutes to find the entrance, but a seedy man in a faded uniform was expecting me. We walked through the building down long corridors to where three civil servants had kindly agreed to break their holiday and tell me about Polish agriculture.

I asked how farmers could possibly manage to survive inflation at 40 per cent per month. The ministry team agreed that the problem was serious, but pointed out that things might not be quite as bad as they first appeared. For example, they told me, fertiliser sales were down 35 per cent this year because the price has risen too steeply. Yet Polish farmers, used to continuous shortages, hoard most of their inputs and today are living off their stocks, safe in the knowledge that supplies are – for once in their lifetime – not at risk. It was an ingenious theory of the sort clever civil servants delight in concocting. And maybe they are right. But it is certainly significant that for the first time the State-owned fertiliser manufacturers have felt the need to advertise on television.

When it came to the structure of Polish agriculture, the civil servants were more frank. 'We've tried to increase the size of our farms for some years now, but without much success,' admitted an important-looking lady who was in charge of Social Affairs. 'But things are changing fast under the new system.' I learned, for example, that it was no longer necessary for a farmer to give his land to the State if he wanted a retirement pension. The old limit on farm sizes of 100 hectares was also about to be abolished, and with it the requirement that you must be a qualified farmer or farmer's son in order to buy agricultural land.

Another civil servant told me, with ill-suppressed pride, that only the day before a farmer from the Poznan area had appeared on television. He owns 80 hectares on which he fattens 2,000 pigs. But what had excited the Ministry man was that the farmer had taken out a loan of £100,000 to put up his own slaughterhouse. 'This man is today a hero, and a wonderful example for all Polish farmers,' I was told. I wondered how he would have been described a year ago if I had been talking to the same civil servant.

As I left Warsaw and headed west towards Poznan, I began to see the German influence reassert itself. The farm buildings were more solid, the farms bigger and, for the first time, I even saw cows grazing grass instead of being zero-grazed indoors.

In Poznan I was invited to share Easter lunch, the most important event of the year, with a local family. It was full of symbolism, most of which I could not understand. The grandfather began by handing round slices of hard-boiled egg which we ate with salt. Then, after handshakes, kisses and elaborate toasts, we began with a cup of borscht before settling down to seven different sorts of sausage, cold ham, beef and stuffed pork. An hour or so later a selection of cakes appeared and more glasses of tea. The language barrier was total, so I sat, smiled and gurgled to show my appreciation. The sentiment was reciprocated when, with great ceremony, I was presented with an illustrated volume of sugar beet diseases. I now know that the Polish for Sclerotinia is Zgnilizna twardzikowa.

Late in the afternoon on Easter Sunday I collected my car from a parking lot in Poznan. The attendant, like most Poles, was keen to practise his English. 'Happy Christmas,' he said with a smile. I smiled back.

> Poland is a bit like Ireland, full of kind, cheerful, Catholic and feckless people who may drink a bit too much but are among the most hospitable humans on earth. Their farming also shows some similarities to Ireland with very small units which somehow defy the laws of economics. I love Poland and the Poles.

ROMANIA
1990

An expensive pedigree sow was bought by a Romanian State Farm. Eventually, the animal farrowed and the stockman was horrified to find only three piglets. He reported this fact to the State Farm Director who, fearing for his job, informed the local Communist Party secretary that the sow had produced a litter of ten. By the time this figure reached Bucharest, it had risen to 20. The Minister of Agriculture thereupon informed President Ceausescu that, in yet another triumph for Socialist Agriculture, the sow had produced no fewer than 25 piglets. Ceausescu decreed that only three should be exported and the remainder kept for domestic consumption.

The story may amuse British farmers, but few Romanians can manage even a smile. Until the revolution last December, the statistics had been faked and the produce exported. As a result, 23 million Romanians endured food shortages which would be rare in the Third World. Even today, with more in the shops than six months ago, many Romanians go to bed hungry each night.

While I was in Romania an echo of the old anger exploded into riots, deaths and what many fear could be a return to the old repression.

In the northern hill country I met Maria. The sunlight flashed off her stainless steel teeth as she greeted me with a V-sign and a cry of 'Victoria'. Like most Romanians, she is still excited by her new freedom. Her red and white cow, Doina, was grazing on the edge of a hay field and, although hobbled, clearly needed Maria to ensure that it did not trespass.

Maria's attitude was typical. Here in the mountain regions, the revolution is still taking place, albeit quietly and without the bloodshed of Bucharest. The effect on agriculture has been cataclysmic. Unlike the rest of Romania – and indeed the rest of Eastern Europe – the cooperative system has started to crumble. Many of the farms have already disappeared.

The land in this part of Romania was never really suited to large-scale farming, and the memories of the old days had been kept alive in the small mountain villages. The cooperative farms up in the hills just disappeared 'like snow in the spring,' I was told.

Two women were turning hay by the roadside at Hidiselu de Jos. The elder, aged 67, used to own 12 hectares in the old days, and has now taken it back from the cooperative. 'I have to work a lot harder than I used to,' she told me with a twinkle, 'but at least I know I will make more money if I work harder, and I know that the money will come to me and not to someone sitting in an office.' The family's livestock consists of one horse, one cow, four pigs and a dozen hens. 'We hope to be able to buy more pigs,' said the younger woman. 'And another cow,' interrupted the elder.

What, I wondered, had happened to the cooperative farms. 'They've gone,' was the terse reply. Then a short pause. 'Well, most of them have, but there are a few people who didn't own any land and so can't reclaim any.'

How did these individuals feel to see their colleagues doing so well? The old lady looked at the ground and the younger one felt she should be spokesperson.

'They are now very unhappy. They feel that they too are entitled to some land because they have also worked for many years. It is because of this that they want to keep the cooperative farms in existence.' The old lady was unable to remain silent. 'Of course these people are in the minority in the village. Anyway, we didn't wait for them to be convinced. We just took our land back again and started farming it.' It was, I reflected, a typically Romanian solution.

The village of Ceica is very lovely. Hills reminiscent of Devon roll steeply away from the white-painted church. Fields of oilseed rape are spattered with brilliant blue cornflowers. The standard of husbandry might be higher but the colours could not be improved.

Until two months ago Ceica was the centre of a cooperative farm which covered 1,600 hectares and employed 600 men. Today the cooperative is derelict. Empty cowsheds lie silent in the sunshine, while in the old farmyard three women sort through a pile of old bricks because they have nothing else to do. Inside the farm offices, where a broody hen was hatching eggs in a cardboard box, I found Ovidu Gavrilu sitting behind an empty desk.

'I do have some work to do,' he told me with a grin. 'I'm also in charge of the local machinery centre here.' He led me out into the sunshine where a collection of rusting tractors and combines looked as if they were awaiting an auctioneer. 'We did some work this spring,' explained Ovidu, 'because the peasants who took their land back needed us to plant their crops.'

Seven men lounged around a combine which most scrap dealers would think twice about buying. 'We've got seven of these,' said Ovidu, 'and we're trying to get them ready for harvest which will begin in less than two weeks.' It seemed a hopeless task.

The peasants in this part of Romania had not waited for the new parliament to legislate; they simply went along to the cooperatives and took their land and animals back. They all knew exactly how many cows, pigs, and sheep they had brought into the cooperative back in 1949 and they allocated lottery tickets for each animal. It was thus a matter of luck whether they got good or bad ones. The remaining animals were allocated to those people who had not brought animals into the cooperative originally.

'Put that tape recorder away,' hissed Gheorge Laza, who directs the Arable Research Centre at Oradea and looks uncannily like Henry Kissinger. 'We've had enough trouble with the Securitate and it makes me nervous. In the old days I had to look under my desk each day to see if I could find a microphone.'

We were high in the mountains eating local cheese tarts called Placinta cu Brinza and drinking local plum brandy. Outside, the rain poured out of a dark grey sky.

Laza went on to explain that he was not as optimistic as some. 'The big mistake we made was that we were in such a hurry after the revolution in December that we didn't have time to consolidate our holdings and now, as a result, the peasants still occupy land which is hopelessly split between many small strips.'

Gaig Vassily was sitting on the ground looking at distant mountains when I approached him from behind. He turned round hurriedly to reveal a brown and crinkled face which broke into a grin when he learnt that I wanted to look at his cattle. Surrounding him on the level plain were a herd of 180 cows and 20 buffalo. It was the latter which interested me. Small tinkling bells, more suited to sheep, had been attached to the necks of these huge, shaggy animals.

The buffalo, explained Vassily, are used both for transport and for milk. 'Buffalo milk is so thick that you can turn a jug of it upside down and nothing will fall out,' joked the ancient herder. I later discovered that it contains 8 per cent butterfat and makes the finest yogurt and cheese.

The mountainous region of northwestern Romania, with its wooden churches and deep valleys, is far from typical. Around Timisoara, where the revolution started, I found myself back on a level plain where vast cooperative farms looked like any others in Eastern Europe. I had been collected that morning by a jeep containing the Regional Director of Agriculture, together with Nicolae Saulescu, Romania's leading wheat breeder, who was my guide throughout Romania.

This is big country where the smallest fields are 100 hectares. Wheat, sunflowers, sugar beet and the occasional pale blue blur of linseed made me feel very much at home. Suddenly, among the oil wells which litter this part of Romania, I noticed a field of bushy green plants which I could not recognise. 'Hemp,' explained Nicolae.

It looked oddly familiar. 'You might know it as cannabis,' explained the wheat breeder. 'We grow it for its fibres which make a fabric a bit like linen.'

But surely, I asked, in what I hoped was a nonchalant manner, people also smoke the leaves? The reaction was one of uncomprehending outrage. 'No. Never.'

There I was, up to my waist in 100 hectares of cannabis. In England a window box of the stuff would send me to jail. 'Of course we breed cannabis for its fibres and not for its other characteristics,' reassured Nicolae. But even he was not entirely convinced by his own argument. 'I suppose,' he mused, 'that we shall probably have to give up the crop in the next few years – particularly if we start to get a lot of tourists.' I was about to agree when he continued, 'But from an agricultural point of view this would be a pity because it's the best possible entry for wheat. And for the local population it wouldn't be good news either because there is a tradition round here that when the hemp is on flower the perfume is highly aphrodisiac.' The Regional Director grinned. 'You should try making love in a field of cannabis when it's on flower. There's nothing like it.'

But it was not just courting couples and wheat growers who found cannabis attractive. It also acts as a superb cleaning crop because it is so thick it smothers all weeds without even needing a herbicide. 'It will even kill couch,' said Nicolae. And as if that were not enough reason to grow the crop, he continued, 'It is so responsive to fertiliser that if you don't apply any to the field the cannabis will give you a map of your soil fertility.'

The total of my agricultural, sociological and sexual knowledge had increased dramatically that morning.

After one of the more intriguing stops on my Eastern European odyssey, our jeep set off again at high speed down the straight but badly surfaced road lined with cherry, damson and walnut trees. We were heading towards the western extremity of Romania where the Hungarian and Yugoslavian borders meet. 'When the cock crows here,' said the director, 'he can be heard in three countries.'

The State Farm at Becicherecuk Mic is anything but a showpiece. Housed in what used to be an old Russian army camp, it was in stark contrast to some of the luxurious offices I had seen elsewhere in Eastern Europe. But if the facilities were tatty, the hospitality was overwhelming. 'Have a bit of breakfast,' said Constantin Buzatu, the young director, as he extracted three packets of Kent cigarettes to give to his important visitors. Kent cigarettes are the alternative black market currency in Romania. We were ushered into a small room where, at 9.30am I was horrified to see a bottle of whiskey sitting among the fried eggs, meatballs, salami, ham, cheese, salad and coffee we were being offered for break-fast. And this in a country where food is still scarce.

Had the recent revolution made any difference to a State Farm, I wondered? Constantin Buzatu thought for a very long time before replying. 'In one sense it's made my life easier, but in another it's more difficult. The good thing is that at last we can get enough fuel, and this has made an enormous difference.' He went on to explain that under the old regime fuel was allocated on a strict formula based on the number of hectares to plough and the crops which were being grown. It was distributed every month by the regional authorities, and was invariably too little. 'Can you imagine,' he continued, 'that we were unable to plough 500 hectares this winter and had to wait until we got more fuel in the spring?'

In addition to a shortage of fuel, there were only four herbicides available, so many of the crops did not receive the correct chemicals. 'We used to say we had to use hoeicides instead of herbicides,' chimed in the technical adviser of the farm. And then, more seriously, he asked whether I had noticed that many of the fields had a lot of weeds. 'What could we do?' he asked. 'We simply don't have enough people to hoe every field by hand. But now we can buy most of the herbicides we need so if you come back next year you'll see the fields looking a lot cleaner.'

So much for the advantages; what about the problems? The director looked at the technical adviser, who looked at the regional boss. Eventually the director began. 'We are having difficulties getting people to work,' he eventually said. 'They now realise that they have rights and are not actually obliged to work, whereas before they had no option.' I probed for more details and the following facts emerged. Under the Ceausescu regime no overtime was paid. Employees on both State and cooperative farms simply worked for as many hours as they were told to. Their wage was a fixed daily rate and bore no relation to the amount they worked, except for a few jobs which were paid as piecework.

'Now,' continued the director, 'it is hard to get them to work more than eight hours a day. And when they do work, we have to pay them for any extra hours and also for weekend work. There are laws controlling these things.' I assumed this was new legislation introduced by the recently elected parliament. 'Oh no,' explained the director, 'the law has always existed but it was simply ignored under the Ceausescu regime.' The regional boss, by then very articulate after three whiskies, interrupted. 'For the first three months after the revolution everybody became politicians. They wanted to argue all day and not to work.'

Had anything else changed? 'Oh yes,' exclaimed the regional director. 'At least we can now tell the truth.' He grinned at the others sitting round the table and began to explain the old system which required every level of management to falsify production figures. 'If Ceausescu had survived for another three years,' he continued, 'Romania would have had the highest yielding crops in the entire world!'

Why had none of the cooperative farms in the Timisoara area been split up amongst the members? 'The land round here demands people,' explained Nicolae Brailoiu, who was by then gulping his third whiskey, 'instead of the people demanding land as they do in the mountains where you were yesterday.' This sounded good but did not seem to make a lot of sense.

The State Farm of St. Nicolae Mare occupies dignified premises in the centre of the town opposite the church. The director's office was notable for the single nail which still protruded from the bare pale green wall behind his desk and which, until six months earlier, would have supported an ornate portrait of President Ceausescu. But times have changed in Romania, and I was interested to know if the director had changed too.

'No. I've worked here for 31 years,' announced Dr. Joan Marcu, 'and I've been director for the past 15.' He certainly looked the part, with pale blue eyes, short-cropped white hair and a profile which suggested he was used to getting his own way. Throughout the conversation he kept walking over to his desk and barking out orders on either the telephone or the farm's radio. After one particularly staccato exchange, I asked what the topic had been. 'I've told them to start the combines even though the moisture is around 22 per cent,' he explained.

As with the previous State Farm, this one had also suffered from a lack of fuel and had been unable to plough five per cent of the acreage last winter. 'You did well,' the regional director told Marcu. 'In Timisoara County the average figure was nearly 30 per cent.'

What about the future? 'For the time being I'm certain that State Farms like this one should remain State-owned, but eventually I don't see why the government shouldn't sell a minority of shares in each farm to the workers. But not,' he added forcefully, 'a majority.'

When it came to the future of cooperative farms, he was equally clear. 'I see no reason why they shouldn't give shares away to their members based on two criteria: how much land the man brought with him into the cooperative, and how long he has worked here. This way even the people who did not own land in the beginning would get some benefit. If you only allocate shares to the people who originally had land, then all you'll be doing is to perpetuate the inequalities which existed 40 years ago, and that would be a terrible mistake.'

Back on his own farm, one thing did worry Joan Marcu. 'The biggest problem we face is with the livestock. Today we are milking 1,000 cows through a single parlour. I agree that the herd is far too big but what can we do? We've got the buildings and we've got the milking parlour. In an ideal world I would like to see four herds of around 250 cows.' How did he feel about his sheep flock? 'I suspect,' he mused, 'that the optimum size of a flock is no more than 2,000 ewes. However, it's all very well my talking like this; the fact remains that it will be hard – even impossible – to achieve.'

It was time for lunch. The farm's jeep deposited us at a modern building on the edge of town. Up a marble staircase to the first floor, I found myself in a small dining room with red velvet panels. It was clearly the Executive Suite. The main topic of conversation was Romania's chances that evening against Cameroon in the World Cup (they lost 2–1). But after two glasses of peach brandy and a lot of local red wine, the assembled throng began to let their political inhibitions loose.

'I know I've been part of the system for the past 15 years,' admitted the director, 'and now we've got a lot of changes to make. But the fact remains that Romanian agriculture has advanced a lot since my father was a farmer with his 15 hectares in the mountains of Transylvania after the war.' It turned out that Joan was the only one of four children to leave his mountain village and get a good education. 'Today,' he continued, 'I am a doctor of agriculture while my brother, who is 63, is still working on the cooperative. My father joined in the beginning.' There was a brief pause before he continued. 'I must admit, however, that today my brother is one of a group of 20 cooperative members who have asked for their land to be returned and are planning to set up their own small group of independent farmers which will work inside the old cooperative.' The rest of us remained silent but wondered how a State Farm director felt about this lack of loyalty by his brother to the old system. None of us dared ask.

Had his father welcomed the collectivisation of agriculture, I asked? 'No. Quite the contrary. In those days just after the war we had four oxen which we used for ploughing, one horse, two cows and a few sheep. Yet in spite of the fact that my father and mother worked every hour God sent them, they hated the idea of collectivisation. In fact my father held out for five years before he finally gave in. During this time he became a sick man with very bad arthritis, but he still refused to give up his independence.'

What sort of pressure did the authorities use? 'The main method was by setting us production targets which were impossibly high,' replied the director. 'We even got to the point of having to go out and buy produce simply to be able to deliver the right amount to the authorities. Then the police used to come and arrest him for no apparent reason.' He paused to refill my glass. 'They would take him away and neither my mother nor the rest of us knew if we'd ever see him again. But then they'd bring him back and hope he'd learned his lesson. It was all designed to frighten him, and in the end I supposed it worked.'

Had this policy affected Joan Marcu at all? 'Oh yes. I spent two years working on a building site because I couldn't get in to university. They refused to let me in because my father had not joined the cooperative farm. It wasn't a pleasant experience for my father or the rest of his family.' It was hard for me to believe that the man who was describing this brutal regime was the same man who had flourished under Ceauşescu's even more tyrannical government.

It was time for the maid to remove the soup and bring on pork fillets. The director refilled my glass with wine before continuing. 'There's no question that collectivisation in this country was forced and not voluntary. But it is also easy to criticise something which happened 45 years ago. As a matter of fact I suspect that most of the peasant farmers did not agree with my father, and were fairly happy with the new system. Where we started to go seriously wrong in this country was in 1978 when Ceausescu decided to export everything we produced and not to allow any imports. It meant that we began to see starvation in our own country, even though we produce a lot of food.'

Topolovatu Mare is an ugly village which straddles the road between Timisoara and Bucharest. In the middle of the main (and only) street stands the headquarters of the local cooperative farm. I walked up the shallow marble steps through swing doors and found myself in an imposing but shabby foyer which felt like the wrong end of the Slough Trading Estate in the late 1940s. Vast ears of maize in a glass case reminded me I was in a farm office. I continued up a narrow curving staircase to the mezzanine floor where a dumpy secretary ushered me through plastic padded doors into an ante-chamber. It was clear that I was nearing the seat of power as the furnishings became ever more garish (the Romanian equivalent of luxurious).

Three grizzled peasants and an old lady wearing a headscarf sat on a plastic sofa waiting to be summoned to The Presence. The president, I was informed, had not arrived, but would I care for breakfast? I readily agreed since I had been on the road for an hour. At a table with a very white cloth, a plate of cold meat immediately appeared, accompanied by Turkish coffee and a glass of peach nectar. Ten minutes later the Great Man arrived. Sporting gold-rimmed glasses, a brown suit, a brown trilby hat, brownish-red shirt and a brown-red tie, he looked like Arthur Daley's Romanian cousin. I felt as if he was going to offer me a good rate for my dollars, or possibly a slightly used ghetto blaster from Taiwan.

After refusing a coffee, 'I have high blood pressure,' Joan Josu invited me to his office. We returned to where the peons were waiting for an audience while Josu fumbled with his keys before unlocking yet another padded door. I had been to at least three dozen cooperative farm directors' offices in my travels round Eastern Europe, but nothing had prepared me for what now appeared. A small room, panelled with dark varnished wood inlaid with elaborate patterns made me feel as if I had stepped into a Mafia leader's den somewhere in Sicily. Red velvet chairs stood on a Turkish carpet. He was proud of his premises and, squeezing my arm tightly, steered me into the boardroom which, with the addition of a long table, was furnished in identical style. Off it lay yet another room in which important business must be transacted; it contained two vast armchairs, two ashtray-spittoons and a single small coffee table.

Back in the throne room, I noticed the bookcase which, in the past, would have been filled with medallions presented by a grateful Ceausescu. Today the shelves were empty, save for an icon of the Virgin Mary and a Bible.

Joan Josu is 56 and has been president of this cooperative for the past 30 years. I learned later that he is one of the best-known names in Romanian agriculture, having been honoured by the old regime for his valiant efforts to promote socialist agriculture. Although he protested that he was entirely happy with the recent revolution, I suspected that he hankered after the old days when he and his henchmen did not have to worry about trivia like democracy.

Did he have any problems with his 300 full-time workers? 'Not at all. My only problem is that we don't have enough people to work on the farm because the younger generation has gone off to work in factories. Before the revolution we couldn't raise the wages of our workers but now this is possible, so all we'll do is to put up the pay and attract them back again.' Very glib. Very confident.

The night before, the riots had started in Bucharest. How did Joan Josu feel about this? 'Well,' he mused, 'I suppose the government should have tried to settle things peacefully, but they couldn't let a bunch of hooligans take over the country, could they?' So repelled was I by his manner and appearance, that I could not believe a word he said. My suspicions were partially confirmed when, a few minutes later, he admitted that on the farm there had been some problems with the political parties who had apparently fomented discontent among the labour force. This must, I reflected, have been most confusing for an old member of the communist establishment who had been brought up on a diet of totalitarianism.

Nobody had asked for his land back and the president wasn't at all surprised. 'They're all happy here. Why should they want to change?' The credo of a true conservative. But he did foresee the day when he would have to distribute shares to all the cooperative members. I was certain that when this day ever did come, Joan Josu would have long ceased to be the farm's president.

As I left, he put his arm round my shoulders, offered me a bottle of the farm's own peach nectar, bid me a fond farewell, and turned wearily back to his office. By then the line of peasants awaiting an audience was stretching out into the hall and down the stairs.

'One of the many problems with Ceausescu,' said Nicolae Saulescu later that morning, 'was that he was particularly interested in agriculture.' I thought this should have been an advantage, but Nicolae explained why I was wrong. 'He used to fly all over the country by helicopter, land by an exceptional field of wheat and then compute the eventual yield by counting grains in a handful of ears. This predicted yield would then be announced as national average wheat yield.' It sounded too lunatic to be serious, but Nicolae continued. 'Ceausescu had a fixed idea that yield would inevitably be increased if the farmers drilled early and used a high seed rate of 300kg per hectare. As a result Romania was the only country in the world where drilling dates and seed rates were laid down by law.' My mind boggled, and then re-boggled.

That afternoon, on my way south through the Carpathians, I achieved one of my many ambitions when I was allowed to try ploughing with an ox. I had stopped the car to watch four peasants hoeing their maize, when I noticed an old lady leading an ox while her husband controlled the plough. Even on a level field it would have been hard to keep the plough straight, but on a side hill I found it totally impossible.

Bucharest was dark when I arrived, and I went straight to my hotel on the outskirts. I am glad I did, because at that precise moment 10,000 miners were running amok in the central district. They had come in from the provinces in response to a call from the president, who found that the police were unable to control a rapidly escalating demonstration by students and opposition groups. The Ministry of the Interior and Television station had already been badly damaged and the government was blaming (with no evident proof) students, opposition parties and groups who had been influenced from abroad.

The miners arrived by train and bus, armed with sledge hammers and pick axes. They set about their job with enthusiasm born of resentment, beating up anybody they suspected of opposing the government. By the time I reached Bucharest, five people had been killed, 1,000 arrested and parts of the city were in a state of armed siege.

That night I dined in the hotel restaurant. As usual, there was no menu and I was informed that the meal would be pork. My conversation with the waiter was unusual, even by Romanian standards.

'I'll have some red wine please.'
'Sorry, we don't have any red wine today.'
'Fine. I'll have beer instead.'
'I'm afraid we have no beer.'
'OK. Then bring some mineral water.'
'We've run out of mineral water.'
'What is there to drink this evening?'
'Champagne.'

So I drank a bottle of warm, sweet champagne with my pork. There is not a lot of food or drink in Romania today.

The following morning I had an appointment at the Ministry of Agriculture, located just off University Square where the fighting had been fiercest. In the pouring rain I found a city with a hangover. People walked numbly through the drizzle without looking to either side. Soldiers and policemen, both armed with machine guns, stood in long rows along the pavement, staring out at the thin traffic. On the walls graffiti and posters gave a hint of the activity which had gone on in the previous weeks. Nearby, burnt-out cars and smashed glass littered the streets. I walked up a tree-lined boulevard to where the headquarters of the National Liberal Party had been wrecked. A solitary soldier stood guard and passers-by pretended not to notice.

It was almost a relief to escape from this dismal scene into the Ministry of Agriculture, where I was met by Ion Mocanu, the Director of Cereals and Industrial Crops. Like all civil servants, he was enormously relieved that Ceausescu had been overthrown. 'You can't imagine how difficult it was for us,' he told me. 'We were forced to report a cereal harvest of 18 million tonnes last year when we all knew perfectly well the figure was only ten-and-a-half million tonnes.' But it wasn't simply the falsification of crop production figures which distressed Ion Mocanu. 'I don't suppose you will believe me when I tell you that last spring this Ministry announced that all spring planting had to be completed by 10 April. They took no notice of weather problems, or the fact that there was not enough tractor fuel. they simply announced this as a requirement.' I was pondering the enormity of such a decree when he continued. 'Of course it meant that all of the farm directors were forced to lie and announce that they had finished planting on 10 April. Any who told the truth and admitted that they had not finished would have been fired.'

But this was not the only stain on Ceausescu's system. What, I wondered, had happened to the idea of 'villagisation' whereby entire villages were bulldozed and 'agro-industrial towns' erected in their place to house farmworkers. Mr. Mocanu looked uneasy. 'It was a terrible mistake, terrible.' There was a silence while he tried not to remember this piece of very recent history. 'More than any other scheme, this did the most damage to our agriculture and caused the peasants to be very unhappy indeed. Because of this the food shortages grew even worse. But,' he muttered unconvincingly, 'it was nothing to do with the Ministry of Agriculture. Now the whole programme has been stopped and it will not be started again.'

He went on to explain that Ceausescu had been obsessed by the fact that while Romania's population was rising in line with his policy, the amount of arable land was declining. He gave orders that this trend must be stopped, and that by 1995 an additional half million hectares of arable land must be in production. One way to achieve this was to ensure that villages occupied less land by piling people on top of each other in high-rise apartments. Hence the villagisation, which had affected almost 300 settlements.

It was a grey, overcast day and my spirits were low. The villages were impoverished and the farming unimpressive. I had just crossed the Danube, heading east to the Black Sea, when the landscape opened up into a vast rolling plain dominated by an East Anglian sky. Scattered flocks grazed the short grass. I stopped the car to look at one which was being tended by a couple with their young son who was on a donkey. As I raised my camera the boy cried out to his mother. 'Help. Will I now go to prison?' This pathetic cry was the clearest legacy of the Ceausescu years I had come across. In those days a camera and a big car meant only one thing: the Securitate, Ceausescu's secret police. I felt awful.

The village of Victoria stands at the edge of one of the remotest regions in Europe, the Danube Delta. This vast tract of land, inaccessible by road, is bigger than Suffolk. Here the Danube, after winding for 1,500 miles from the Black Forest, finally meets the sea among the marshes and reedbeds which provide a home for over 300 varieties of birds.

I had stopped in Victoria to investigate why a group of people was standing round a small lorry at the crossroads. It turned out to be the milk tanker making its daily collection from the sheep farmers of the district. The local producers were waiting with their churns to have them weighed and then tipped into the top of the tank.

I looked into the tank and wished I had not. Brown scum seemed to be floating on the surface. The small group had between them given 50 litres of milk that afternoon. My eye was immediately attracted by Constanza Auhum who, with another of those stainless steely smiles, and an armful of red roses, was evidently the leader. She described herself as 'an assistant shepherd' on the cooperative. Did she own any land? 'Yes, around three hectares but I'm not very sure really.' Had she thought of asking for the land back again? 'No, why should I? First of all I'm too old, and where would I find the money? Besides, we don't have a bad life on the cooperative. It's a lot better than it was before the revolution.'

At that point Nicolae Toma, who works as an agricultural labourer on the cooperative farm, chimed in. 'You can say that again. We've been given half a hectare each as an allotment, and our wages have gone up quite a lot too. But best of all, we've been given four kilos of grain per day which we never had before, and that comes in useful if, like me, you're fattening pigs.'

Had anybody asked for his land back? 'Yes,' said Constanza, 'two people from the village did but they're crazy. It won't work and they'll soon regret it.' At this point a man sidled up to our group and became more than a little belligerent. 'Don't speak to that person,' he told the crowd, referring to me. 'Get him to tell you who he is and let's see his credentials.' The rest of the group was embarrassed. 'Shut up,' Constanza told him. 'This isn't the old days. We're free now,' and then, turning to me in a whisper, she said, 'Pay no attention. He's always like that. He's probably been drinking.'

The scenery south of Constanta resembles North Dakota, with its vast landscapes and fields which appear to have no ending. Called the Dobrudja (beyond the Danube), it is an arid part of the world with a rainfall of around 350mm per year, but it has superb black chernozem soil. 'Chocolate chernozem' said one farmer. 'It's the caviar of soils' said another. Irrigation channels carry water from the Danube, enabling wheat, maize, sunflowers and soya beans to be grown.

The village of Topraisar is like any other in the Dobrudja except that, being near the Bulgarian border, it has a military camp. The State Farm office has a house martin's nest above the front door. Inside was a list of the contributors to and borrowers from the farm's own Savings Bank. From it one could see the totals of every member's deposits and borrowings. Not the sort of information which would be publicly available in Britain but, as Bomritu Flora, the director, told me. 'It's their money and they have a right to see where it is going.'

High behind Flora's desk protruded the usual solitary nail. I pointed this out and he laughed. 'I'm going to put an Orthodox calendar there,' he told me, 'so I can keep track of the Orthodox holidays. They're different to ours.' He went on to say that there is also a practising mosque in the village, serving the Turks who live in the area.

We tried to go round the farm in the director's Dacia, but the last three days' rain meant that most of the tracks were impassable. 'The revolution has made my life a lot easier,' he told me, 'but there are still a lot of alterations which need making. For example I hope they'll get rid of the centralised planning which Ceausescu loved so much.' I asked for an example, and received one instantly. 'We grow hybrid maize for Pioneer Seeds in America and they, of course, pay in dollars. The snag is that we never see any of those dollars here on the farm. They go to the central marketing agency in Bucharest. If only I could deal direct with Pioneer, I'd keep the dollars myself and buy better equipment.' Flora looked sadly at a rusting engine which was, in theory anyway, meant to pump irrigation water from the channel. 'Look at it. How can we manage to grow half-decent crops with junk like that?' It was hard to disagree.

Dumitru Manole is a small bear of a man, with a voice deep enough to make Lee Marvin sound like a tenor. 'Hi,' he growled when I walked into his office. 'Wanna beer or will a coffee do?' A State Farm director who spoke English, albeit with a strong American accent? It turned out that Dumitru had spent a year in the USA back in 1970 and, thanks to the BBC and occasional visits from Pioneer Seeds technicians, he had been able to keep up his skill. He turned out to be a State Farm Director responsible for 8,000 hectares of irrigated black chernozem soil just north of the Bulgarian border near the Black Sea coast.

Like many good farmers Dumitru was bursting with enthusiasm. 'Come and take a look at what we do here,' he said as he squeezed my arm painfully (Romanians are great arm-squeezers, back-patters and hand-holders). We piled into a pick-up truck and went bumping through the puddles to where 12 combines were just starting this year's harvest on a field of barley. The number of machines was impressive, but nothing else about them was. Dumitru could read my thoughts. 'We have to make do with what we've got,' he continued, 'because there just aren't any spare parts available. I can't find a battery for love or money round here, and the position with belts, tyres and bearings is just as bad.'

It was clear he found it hard to show this side of his farm to a visiting foreigner. After a moment he snapped out of his depression. 'Just wait,' he said. 'If we can only get good equipment here, we'll be able to beat the world with our yields. But,' he added, 'I don't just mean combines and tractors. What I really want is good irrigation equipment. We're still using ancient pipes which involve a lot of hand labour and cost a fortune. The pumps are worn out and the sprinklers are knackered.'

There was a pause while he tried to restrain his bitterness. 'Give me good irrigation equipment and I'll show you what could be done in this part of Romania. You'd be amazed. We've got the soil, we've got the climate and we've got the water. But thanks to Ceausescu, we've lost 20 years and it's going to take a long time to catch up.' Dumitru Manole had just summed up Romanian agriculture. There was nothing left to say.

Romania was a tragedy and, I suspect, still is. The land – especially that by the Black Sea – is magnificent, the climate superb. The potential is, as my land agent friends would doubtless say, limitless. And yet it never quite seems to work out.

BULGARIA
1990

Umpteen centuries ago, in some remote and filthy Bulgarian cowshed, the milk went sour. The little bug which caused the problem was later named *Lactobacillus Bulgaricus*, and the sour milk is today called yogurt.

In the canteen of the Institute for Wheat and Sunflowers outside the town of General Toshevo I had my first taste of the real stuff. It is what yogurt once was but never is. Smooth, rich, slightly sour and very delicate. I thought I had died and gone to heaven.

Minko Stoyanov's small office looked a very long way from heaven. There were cigarette ends on the floor, empty coffee cups from an earlier meeting were scattered around and one of the windows was broken. Forgoing the panelled walls and plush furniture of other Cooperative Farm Presidents, Stoyanov manages to survive with a desk, sofa and empty book case. The only luxury was an electric shoe-polishing machine of the sort you find in cheap British hotels.

Stoyanov does not look like a farm manager. His lank, grey hair and pudgy, expressionless face reminded me of an accountant I once knew. After a few minutes of pleasantries, we were joined by three of his lieutenants, who, having squeezed onto the single small sofa, proceeded to nod vigorously at what they hoped were the right moments. Longing for some fresh air, I suggested we might look round the farm, and was driven by the President in his Lada while the lieutenants followed at a respectful distance in a jeep.

The Stefanavo cooperative prides itself at growing some of the best wheat in the country. Located in northeast Bulgaria, the soil is superb black chernozem one metre deep and the countryside is bespattered with trees. These, it seemed, are the remains of a vast forest which was cut down in the early 1960s.

In addition to wheat, the cooperative grows magnificent sunflowers. They were tall, dark green and even, with the first flowers just beginning to appear. I was not surprised to learn that one third of the French sunflower crop is planted with Bulgarian varieties. The farm also produces maize, sugar beet, soya beans and French beans which are harvested dry.

The scale of the operation was bigger than anything I had experienced in the rest of Eastern Europe, with some fields of more than 250 hectares. As we drove round the farm in the late evening sun, I saw herds of deer and several partridges. The forests are full of wild boar and in the villages storks nest on chimneys and electricity poles.

Back at the main farm buildings, I was surprised to see a few workers still tinkering with their machines. The reason became clear when I realised that they were working on four new red Russian Don 1500 combines. Like all farmers, the President was eager to show off his toys. 'They only arrived this afternoon,' he explained, 'so we haven't had time to clean them up.'

Kolyo Marchev is aged 50. When I met him he was greasing a bearing on one of the machines. 'They're a lot better than the old ones,' he told me. 'For a start they're all automatic these days, and they've got air conditioning which makes harvest a lot easier. We get temperatures up to 40 degrees in August.'

It turned out that Kolyo's father used to own eight hectares which went into the cooperative when it started in 1947. 'I've worked here all my life,' he told me, 'and I wouldn't want to change.' How about setting up again as a private farmer on the ancestral eight hectares? 'Why should I?' came the reply. 'I wouldn't gain anything. In fact I'd be worse off, and I certainly wouldn't have a machine like this to drive, would I? Anyway, I can't remember what it was like when my father farmed the land. I was too young then. All I know is that he and my mother worked damned hard.'

It transpired that Kolyo Marchev was, in a sense, already a private farmer. 'I've got five sheep, five pigs, a hundred chickens and a donkey called Mara,' he told me. Last month he had sold two of his pigs to the State. 'The money comes in handy, but that isn't the only reason I keep these animals,' he added. 'I just like the life. Besides, it's what we all do in this village. In fact it would be unthinkable not to have a few animals.'

I was mildly surprised to hear that, like all farmworkers, Kolyo was not paid extra for overtime. 'But we get a share of the profit at the end of the year, so we really benefit,' he reassured me.

We went out to look at the dairy herd. Stoyanov explained that four years ago they had bought 600 in-calf Friesian heifers from France, which were housed in three huge buildings. Milking was in progress and I watched while women moved slowly up and down the standings, plugging the units into the vacuum line. The livestock manager, Dimitri Dimov, joined us and explained that the herd has been divided into groups of 50 animals, each of which is looked after by one person, so the cattle get individual attention.

I wondered what happened to the bull calves. 'We take them up to 150 kilos and then sell them to members of the cooperative for 3.50 lev per kilo. We then buy them back at the same price when they weigh 550 kilos,' explained Stoyanov. This enables the workers to supplement their income by fattening cattle without having to invest too much money.

The conversation turned, inevitably, to politics. Had any members, I wondered, asked for their land back? 'Yes,' replied Stoyanov, 'two people out of 380 have asked me, but I told them we would have to wait and see. No decision about land ownership has yet been taken by the new government, but I'm sure this will be one of the first things they will think about when the new parliament meets this month.'

Had the change in government – it has hardly been a revolution – made any difference to a cooperative farm president? 'Oh yes. Enormous,' was the reply. 'I've been working here for the past 32 years, and this is the first time we have been allowed to plan the cropping on our own farm. Until this year we used to receive orders from Sofia telling us what acreages to plant of each crop. And they set us production targets too. But now they have changed all this and they let us decide everything.'

Given this profound change, what alterations would the president make to the farm? He thought for a moment and then had a hurried consultation with his colleagues before continuing, 'Not many really. We'll certainly grow a bit more sugar beet next year because the price has more than doubled. But we're only talking of 100 hectares extra so it won't make a big difference.'

Bulgarian politicians start work early. It was 7.30am when I knocked on Pancho Redev's door in a leafy street just off the main square of Tolbuhin. The hall was bare except for the inevitable shoe-polishing machine. There was a brief wait before a large mustachioed man in a dark suit welcomed me into his book-lined office. Mr. Redev is the chairman of the local Agrarian Party, one of the three main political groupings in Bulgaria today. The previous week it had fought a general election but had not done well. 'We only managed to get eight per cent of the national vote,' the chairman explained, 'but we did slightly better in Tolbuhin.'

Half an hour later, after two cups of Turkish coffee and many explanations, I was no less confused by the Bulgarian political scene. The Socialists, who nar-rowly won the election, are the former Communist Party under a new name. The Democratic Union occupies the right wing and the Agrarian Party, Mr. Redev assured me, is in the centre. So far so clear. But when he came to explain that there are, in fact, two agrarian parties which had split some 50 years ago and had since been allies (even puppets) of the Communists, I began to sympathise with the Bulgarian voters.

It was a relief to get back to farming. Georgi Kolev Georgiev runs the cooperative farm at Donchevo. He looked as if he had been one of the famous Bulgarian weightlifters. Short and squat, his neck was only slightly thinner than his waist. Instead of Todor Zhivkov, the former Communist party boss who is now in jail, a sexy stewardess simpered down from a Bulgarian Airlines calendar. Around the walls sat the farm's top brass who had come to look at an English farmer and hear their boss being questioned.

But before I started my normal interrogation, Georgi Georgiev announced that he wished to ask me a question. I assumed it would be the normal query I had encountered in Eastern Europe: how much did my car cost? But it was more profound than that. 'Can you,' asked the president, 'change your crops in England if the price goes down?' For a moment I was tempted to tell him that we too were ordered what to plant by the Ministry of Agriculture, but I rejected this line in favour of the truth (or a rough approximation thereof). After my brief seminar on the CAP, it became clear that they were as confused by the EEC price supports as I had been by Bulgarian politics.

My meeting with the Agrarian Party was still fresh in my brain so I took a straw poll of the assembled group of five farm managers. Not one supported the Agrarian Party. 'I was happy with the election last week,' said Georgiev. 'Not simply because the best party won, but because it was all so calm. We didn't even have a single bottle thrown in the village.'

'Unlike Romania,' sneered one of his colleagues.

When it came to land reform, the group was unanimous in its opposition to any suggestion that the land should be divided and returned to the original owners. Georgiev emphasised this point by announcing that nobody on the entire farm had expressed a wish to have their land back again. 'They know what is good for them,' he said with a hint of smugness in his voice. 'They know it would mean more work and less money.' He waited for me to digest this pronouncement before continuing. 'I know what I'm talking about because my father was a peasant and I remember what life was like. There were four children and we all had to help out on the farm, which consisted of eight hectares. We had two horses, two cows, four pigs and thirty sheep. And I can tell you it was hard work. Mind you,' he paused to light another acrid cigarette, 'my father managed to make enough money to pay for all his kids to get a good education. And today I'm the only one who's still in agriculture.'

How had his father reacted to the forced collectivisation of his farm? Georgiev appeared not to hear my question, and went on about the importance of education. I repeated the question and he looked slightly uncomfortable. After another pause he coughed. 'I must tell the truth,' he said. 'It was a very difficult period indeed for people like my father.' This brief reply was a coded admission that his father, like all peasants, had loathed collectivisation and had fought against it as much as he could.

It was ironic that a man whose family had been so hurt by the system was now trying to defend it. But defend it he did. 'I've been president of this farm for 15 years and I know how everyone thinks round here. There isn't a chance that anyone would want to return to the conditions which existed after the war. We had no tractors, no mains water and no electricity then.' It was time to drop the subject, but I was left with the strong, and slightly cynical, impression that Georgiev, like most cooperative farm managers, knew he would be out of a job if the land were ever returned to the original owners.

When it came to economics rather than politics, Georgi Kolev Georgiev was a raving radical. Inside this tough Bulgarian communist, a capitalist was trying (and nearly succeeding) to get out. 'Now that we have been freed,' bubbled Georgiev, 'we'll be able to set up businesses on our own here. I think we'll start by building a bakery so we can use our own wheat. Then it would make sense for us to take the milk we produce from the cows and sheep and turn it into cheese or yogurt. We are very close to the town of Tolbuhin with a population of over 100,000, so it would make sense wouldn't it?' I agreed with what turned out to be excessive enthusiasm. 'Perhaps,' continued Georgiev, 'you would be able to help us with the money. Perhaps you know of English companies who would like to join us in these ventures.'

This was rather more than I had bargained for, and I mumbled something about these things being rather complicated. 'But surely you have banks in England?' chimed in a silent man whom I took to be the accountant. 'We have heard a lot about the banks in western Europe and we understand they lend money for projects like we have in mind.' Once again I tried to explain the mysteries of the western system, particularly the role of the clearing and merchant banks in England. Once again I realised that my knowledge was inadequate.

The audience remained unconvinced. 'But,' announced a thin, bald man who had been chain-smoking nervously, 'you say that you write for a British farming magazine. Why could you not advertise our farm so that other people will know about us?' It seemed a reasonable – if slightly naive – question. I trust the preceding paragraphs have done just that.

The farm's minibus was waiting for me, and in it sat Katya Davidova, a cheerful, shy and well-insulated lady of 30 who is in charge of the sheep operation. She told me that lambing begins in early November and finishes at Christmas. According to Katya, they get 1.1 lambs per ewe, but since I have yet to find a farmer who tells the truth about lambing percentages, I automatically (and possibly unfairly) adjusted this figure downwards.

We took a brief look at the buildings, where a few scraggy Merinos still lurked, before setting off up a steep hillside to where the farm's 3,200 ewes are grazed on unploughable land. At the top of the hill I got out of the van into warm sunshine, where the smell of herbs and wild flowers was intoxicating. Vivid purple thistles grew waist-high, but the rest of the grass seemed unexceptional. Only when I looked closely did I notice a host of plants I did not recognise. These, I can only assume, provided the magical smell which I shall remember for a very long time.

In the distance small groups of 250 ewes were being watched over by shepherds, each of whom had a crook but no dog. In spite of this, however, he was able to control his flock by making clucking noises which the sheep understood perfectly. 'We used to have sheepdogs,' said Katya, 'but a few years ago the government decided that dogs carried diseases and they were forbidden. The shepherds had to learn alternative techniques.' And so they had. Throughout the whole of Bulgaria, where there are lots of sheep, I noticed only a single sheepdog.

In the bookshelf behind Ivan Baiev were the collected works of Lenin – all 55 volumes. This was particularly odd since Baiev assured me that he had just voted for the Agrarian Party. What made it still more strange was that, like the other cooperative farm presidents I had met, he was implacably opposed to land reform – a main plank in the Agrarian Party's platform. He seemed to be hedging his political bets more than most.

His three colleagues were still members of the Communist Party which had recently changed its name (and even its policies). Did this, I wondered, make life awkward in the top echelons of the Sokolova Cooperative farm. 'Not at all,' laughed Baiev, 'we all agree about management and we don't really talk politics.'

To escape from the stuffy office, and the ubiquitous shoe-polishing machine, we walked round the farmyard where the combines were being overhauled before harvest. Each of the tractors had branches attached to the roof, making the farm's fleet look like a mechanised version of Macbeth's Birnham Wood. The reason, however, was practical and not decorative. Each machine must be equipped with a brush to beat out the field fires which are a constant danger in this hot, dry climate.

Elsewhere in the yard, small foals teetered among the tractors while their mothers, who were attached to carts, waited to be shod. The farm owns 30 horses, and the same number of tractors.

'It's going to be a lot easier this year,' said Baiev, 'because for the first time we won't have those people from Sofia interfering as they've done in the past.' It later transpired that people from the Ministry of Agriculture – and the regional Communist Party – used to tell the farms when to start harvest. 'They would order us to begin on a certain date, and we couldn't argue with them,' continued the President. 'It wasn't just harvest either. Each year I would be told when to start drilling – and when to finish.'

I wondered whether the farm was not overmanned. Once again I was informed that the fault lay with central government. 'We were ordered,' explained Baiev, 'to employ two tractor drivers for every tractor we owned.' Whether or not this was a genuine complaint was impossible to determine. It is certainly handy to have an old and discredited regime on which to blame all the ills of the present.

The changes which have resulted from the overthrow of the Zhivkov regime last year are profound. Each farm will be given a quota for so many tonnes of each crop. The government contracts to buy this amount at a fixed price, but any surplus may now be sold by the individual farms on the open market. This has given scope for entrepreneurial managers to flourish in a way that had previously been impossible. 'We've already sold some 400 sheep to people in the area,' said Baiev. 'The odd thing is that while they like to have the animals, they don't seem at all keen to have land themselves. Perhaps it's too much like hard work.' And he chortled at this suggestion. Later that afternoon I saw the first signs of capitalism as a roadside market selling poultry flourished in the shade of a large tree.

The Balkan mountains look better in an atlas than they do in real life. For a range almost as famous as the alps, they were a bit of an anti-climax. I had left the vast rolling landscape of the Dobrudja, and headed southwest into the centre of Bulgaria.

The village of Presvite Kozma is 1,500 feet above sea level. I stopped by the village green to count the different species of animals which were inhabiting the small patch of grass. From the car I could see one mule, several horses, chickens, ducks, geese, two pigs, a lot of sheep and a few goats. In the centre of the green was a well, beside which an ancient Hungarian-type crane had a bucket dangling from the long arm. Two young men, both with transistor radios blaring different programmes, were sitting on the edge of the well.

The younger of the two, a thickset man with a naked lady tattooed on his forearm, was called Stoyan Simenonov. He was a talkative 26 year old who turned out to be a private shepherd. In the four years after completing his National Service, he had built up the flock to 120 ewes.

The statistic, however, of which he was most proud appeared to be that he made 40,000 leva last year. I later discovered that this was double the salary of a university professor and almost four times higher than that of a cooperative farm tractor driver. In Bulgaria nobody hesitates to discuss their salary, and it became immediately clear that Stoyan was well on the way to becoming the Donald Trump of Presvite Kozma. This was confirmed when he told me that many of the other villagers resented him. 'They say I'm making too much money,' he grinned.

I remembered what another Bulgarian had observed earlier. 'We are all swimming in muddy waters these days. Most fish don't like muddy waters but a few, like pike, manage to survive and even flourish.' I had jut met a small pike swimming in splendidly muddy water.

Had there been any pressure on Stoyan to join the local cooperative farm? 'Yes. It wasn't easy for the first few years,' he told me. 'But my father is also a private farmer – he fattens 100 pigs a year – and he helped me to be brave.' I was still unclear why courage had been necessary, and asked Stoyan to explain. 'I used to be hauled into the police station once a week,' he said, 'and they'd try to get heavy with me. Towards the end they even threatened to send me off to another part of Bulgaria. Of course, I didn't take any notice, and nothing happened.' He went on to explain that since the coup on 10 November last year, the police had not bothered him.

Stoyan had taken advantage of this new peace to buy 12 dekar of land (0.12 hectares) on which to graze his 150 lambs. It seemed like an impossible stocking rate. And so it would have been in Britain. Stoyan explained. 'Of course I also use common land like this grass here, as well as roadside verges. But it wouldn't be possible if I wasn't able to buy grain to fatten the lambs.' Some of this grain, it turned out, was part of the payment he received when he sold the fat lambs to the State. 'They love my lambs,' continued Stoyan, 'because they're of a much higher quality than those from the cooperative farms round here. I'm told that the government exports my lambs to the Arabs who like to eat sheep very much.'

Lack of confidence was not one of Stoyan's problems. 'There's nothing to it,' he continued. 'The lambs go to the State when they weigh 35 kilos and they pay me 3.20 leva (25p) per kilo. In fact they've just put the price up to 4.2 leva, so I should do even better this year.'

Given his apparent success, was he worried that his friends and neighbours (who were evidently not the same people) would follow his example and set up on their own? At this point the silent shepherd, a slender youth called Casimir Vladimirov, spoke. 'There's no chance of that. It's too much like hard work for those people. They're not used to work any more because they just lounge about all day on the cooperative farm sitting on tractors or watching cattle chew the cud.'

Casimir, who had been a schoolmate of Stolyan, had a somewhat smaller livestock operation, consisting of only 70 ewes and a goat which was exceedingly affectionate. 'We're thinking of setting up a company,' he told me, 'which will enable us to sell our meat direct to other people without having to go through the government buying agency.' I nodded approvingly, and once again found that I was being regarded as an expert in all forms of capitalism. Would someone in England like to buy their lambs? How do you set up companies in England? Who should they speak to here in Bulgaria? I was forced to admit that, as an arable farmer, I was completely ignorant in matters ovine. 'A farmer without animals?' echoed Stolyan, 'How is this possible?'

Our conversation had been interrupted by enormous explosions which were the outward and audible signs of the nearby Bulgarian army on manoeuvres. At one particularly loud bang, I winced. 'Don't worry,' said Casimir, 'they are very bad at shooting and will not hit you.' It was precisely because I feared that they were bad at shooting that I felt endangered.

Before I left Presvite Kozma, I learned from Casimir and Stolyan about a problem which looms large in Bulgaria today: the Turks. The two shepherds were among the very few Bulgarians in a village with an enormous Turkish majority. Behind me an old Turkish lady, clad in baggy trousers, was painting her wall, and on the other side of the green another Turkish lady was feeding her seven chickens. 'It wasn't too bad until the Zhivkov regime was overthrown,' said Casimir. 'But since then they have become very uppity. Now they are demanding their own school and their own local authority. They listen to Turkish radio the whole time and are turning very militant.' It was a side of Bulgarian politics I did not wish to become embroiled in, so I bid the two private shepherds farewell and, like a hardened war correspondent, drove my car towards the sound of gunfire.

Bjala Cerkva, which means White Church, doesn't actually have one. At least not an active one. The village church closed in 1960 and has not functioned since. But, like so many aspects of Bulgarian life today, this may be about to change. A few weeks ago the authorities allowed the doors to be unlocked and, as a result, a few of the devout villagers now say their prayers in a church they have not entered for 30 years. I had been brought here by Ivan Panayatov, Bulgaria's leading wheat breeder, who speaks English and Japanese and distils fearsome apricot brandy.

Like most small Bulgarian towns, Bjala Cerkva consists of a wide main street on which donkeys outnumber both cars and tractors. Off it run small unpaved alleys where most of the inhabitants live in crumbling houses which, like the church, have not been repaired for ages.

In one of these houses lived Ivan Panayatov's old father. A 75-year-old widower, he is still employed by the cooperative farm as a fire-watcher. Like most cooperative farm members, Mr. Panayatov 'contributed' his own land when Bulgarian agriculture was forcibly collectivised in the late 1940s. Today he lives in his old house, surrounded by a small farmyard in which he keeps three sheep, a few chickens and a donkey to pull his small yellow cart. The donkey had foaled a few days earlier and was in a highly protective mood, kicking me as I entered the dark stable.

One of the reasons why Mr. Panayatov is still employed by the cooperative farm is that he has been friends with the farm president for almost 40 years. As I waited in the outer office to meet Bjala Cervka's leading farmer, it became clear that this cooperative had not moved with the times very fast. Outside, a poster attacking the Democratic Alliance's policy of returning land to the peasants was prominently displayed, and indoors a red flag promoting the (now defunct) Bulgarian Communist Party was on the wall. It was accompanied by a photograph of 35 dismal men who turned out to be the founding fathers of the cooperative back in 1947. No wonder they looked so miserable; they were celebrating the extinction of their own farms.

Eventually the President, Russi Shankov, arrived and ushered me into his office for a coffee and fizzy green bilious lemonade. Behind the desk I noticed something which seemed to sum up Bulgaria perfectly. Lenin, who for the past four decades had stared out from the place of honour in the presidential office, had recently been taken down. But he had not yet disappeared. The only remaining decorations were a Renoir nude and a calendar from the local insurance company.

What would happen to the portrait of the great man, I wondered? 'He'll go into storage in case he is ever needed again,' said Mr. Shankov with a grin which showed it was meant to be a joke (but also concealed some sadness). 'The management committee decided last month that the farm should henceforth be run on completely apolitical lines and I thought it sensible to remove Lenin. But,' here he paused to light a cigarette, 'we should not forget that Lenin also had some very good ideas. It's too easy just to say that everything he did was wrong.'

As he spoke he was searching his desk drawers and finally found what he had been looking for. 'Here,' he announced triumphantly, 'if you like him so much, you can have him.' And he presented me with a small bust of Vladimir Illicit Lenin. It is my favourite souvenir of Bulgaria.

Yet in spite of his nostalgia for the old order, Shankov was obviously enjoying his new freedom from centralised control. 'We can now grow what we like,' he told me happily. 'For example, last year we were ordered to produce 1,500 tonnes of tomatoes. This year we'll only produce 500 tonnes because we simply don't have the labour to cope with this crop. And what is even better, nobody in Sofia can do anything about it.'

I could understand why farm managers enjoyed their new independence, but did it make any difference to the farmworkers themselves? Mr. Shankov had no doubts. 'Of course,' he said with slight aggression. 'They're now getting more money. We are paying them 20 per cent more than last year, and we decided to do this on our own initiative. It would have been impossible until 10 November.' It also transpired that the farm has recently doubled the amount of land that each family can use for growing crops and keeping animals.

I decided to ask Mr. Shankov a hypothetical question. What, I wondered, would he say if the phone rang and the new Minister of Agriculture was asking for his advice?

'I would suggest he did four things immediately,' said Shankov, who had obviously been thinking about these problems for some time. 'First and most important, I would ask him to clear up the whole problem of land and private ownership. Nobody knows what is happening or what is going to happen. Here I am being asked by my members for some land and what can I do? I can't make any decision until the parliament in Sofia makes up its mind. It's becoming impossible.' Mr. Shankov clearly did not like swimming in muddy water.

His second piece of advice would have been as popular in Britain as Bulgaria. 'I would tell him to raise the prices for agricultural produce and lower the taxes we have to pay.' The cry of farmers down the ages. He was particularly annoyed by the fact that when the cooperative farm sells water melons to the state it receives 0.2 leva per kilo, but when a private individual sells water melons in the market place he receives a price five times higher. As a result of this, the cooperative farm had stopped growing water melons.

The laws of supply and demand were already beginning to function in Bulgaria.

Shankov was almost as outraged by the fact that the farm paid ten per cent of its income in tax, but admitted that this was half what the old regime had levied.

I decided not to tell him that by most standards he was extremely well off.

The third reform necessary, continued Shankov, was for cooperative farms to be given the freedom to set their own pay scales.

'My job is getting more and more impossible,' continued Shankov. 'The youngsters who are bright and get good qualifications don't want to stay in farming; they get jobs in industry instead. In recent months, now that the borders have been opened, things have got out of hand completely. Bulgaria has lost over 300,000 people in the last year to the western countries. Take our vet, for example. He's 40 years old and has decided to emigrate to America. He's leaving at the end of the month and there's nothing we can do to stop him.'

The final piece of advice to the Minister of Agriculture was simply a plea for more and better machinery. And this from one of the very rare Bulgarian farm managers who actually runs a Claas combine and, as a result, is regarded with envy by his peers.

The picture which emerged of Bulgarian agriculture was not a happy one, but I had no difficulty believing it. Earlier that morning I had gone to the main post office in Veliko Tarnovo, to buy stamps. It was 10am but the lady behind the counter announced that she was going off for a rest and I would have to wait until she returned. How long would that be? The reply was a shrug of the shoulders. Of such stuff is the Bulgarian economy today.

Lunch in the cavernous and faintly squalid restaurant on Bjala Cerkva's main square was far from memorable. As usual in Bulgaria, the air was heavy with cigarette smoke. There was neither beer nor mineral water and – amazingly – the yogurt had also run out. So I contented myself with a modest plate of cabbage salad and some bread. As I was about to leave, I was accosted by a grizzled individual who, with four empty beer bottles in front of him, was extremely articulate. He introduced himself as Angel Todorof and said that he worked in the local brick factory. 'If you want any bricks, let me know. I'll get them for you cheap.' I thanked him and promised to remember this fact if ever I built my retirement cottage in the hills of central Bulgaria.

When Angel learned that I was English he became still more animated. 'Please send us your tractors,' he said. 'I would very much like to be a farmer myself but I do not have a tractor and the cooperative farm will not sell me one. Besides, they are too big. So why don't English farmers send us their old tractors?' Why indeed?

If Bulgaria is famous for anything today it is roses. From these come attar of roses, essence of roses, rosewater and hard currency. The centre of the rose industry is in the romantically named Valley of Roses. Anticipating a steep alpine valley with roses growing on either side, I drove 200 miles to find myself in a flat plain between two ranges of the Balkans. There was hardly a rose in sight, but perhaps this was unfair as the harvest had taken place a month earlier. Instead there was a lot of wheat and a few fields of purple lavender. Only when I lunched in a dingy roadside snack bar did I actually clap eyes (and nose) on roses. They were drying in piles on the pavement and I picked up a handful. The smell was more intense than any rose I had smelled before. My nose puckered under the weight of the perfume. Subtle it wasn't; lovely it was.

But all good smells must come to an end. It was time to leave Bulgaria, its roses, its yogurt, its shepherds without sheepdogs and its electric shoe-polishing machines. Lenin may still tower over downtown Sofia but a capitalist shepherd high in the Balkans can today make good profits. Bulgarian waters are still very muddy indeed.

I had come to the end of a journey which had taken me 10,000 miles from the Baltic to the Black Sea. Except in Poland, where private farming survived throughout the communist years, the structure of agriculture had been identical. Thus the appearance of farms, whether outside Berlin, Budapest or Bucharest, was uncannily similar. Directors' offices with their padded doors, yards full of rusty combines, cubicle sheds containing six hundred cows and, above all, treeless fields which stretched to the horizon and beyond. The soil and climate might vary, and the skill of the managers certainly did, but the dead hand of centralised planning was always present.

None of this surprised me. It was, after all, what we in the capitalist West had always been told. But one fundamental fact did come as a total shock. The peasants of Eastern Europe, with the exception of the mountain areas of Romania, are not, repeat not, longing to get their land back.

The men who were forced to give up their farms 40 years ago have largely died or retired, but even those few still working today have no desire to turn back the clock and return to a life of unrelieved drudgery on an acreage which they realise is far too small in 1990. Their sons and grandsons are even less enthusiastic about the attractions of independent peasanthood.

So much for my theory that man's desire to own and farm land is a universal imperative exceeded only by his sex drive.

The reasons for this phenomenon are, with hindsight anyway, simple. Peasants the world over are a conservative bunch, frightened by change and content with the status quo. And after 40 years in Eastern Europe the status quo is the collective system of farming. It is no accident that in the recent Bulgarian elections the socialists (the reborn communists) did well in the rural areas while the right wing Democratic Alliance won in the cities.

And yet strong though these attitudes are in 1990, they will weaken. Today the Eastern European peasant regards his family's land in the cooperative as nothing more than a patch of soil which grows crops. But tomorrow this view will change. It will change when other people start paying money for land, either to farm or to build houses, roads and factories. At that moment the peasant will begin to realise that his land is more than a bit of earth; it is also an asset which can be converted into cash. He will then start pressing for the title deeds to be taken out of the cooperative president's safe and, once again, be placed under the mattress of the double bed.

This process will take time, certainly a decade. During this period the whole structure of Eastern European agriculture will also change as modern machinery, techniques and inputs become widespread.

Today most of these countries are net exporters of agricultural products, so it is inevitable that these surpluses will grow still larger. Yet Western European farmers need not panic. Although Comecon, in its present form, will have long since disappeared, the Soviet Union will remain Eastern Europe's natural market for food supplies. Geography and tradition will ensure that the vast majority of Eastern Europe's agricultural produce will continue to move east rather than west.

Nevertheless, on a local level, almost every farm manager I met was keen to make contact with outlets in the west. There is certainly potential for greater exports of fruit, vegetables and, in the case of Bulgaria, dairy products like yogurt.

In the meanwhile, British farmers should not become hysterical. No tidal waves of milk and corn will flood the EEC and bankrupt Cheshire and Cambridgeshire. The farmers of Poland, Czechoslovakia, Hungary, Romania and Bulgaria have a long way to go rebuilding their agriculture – and recovering from a nightmare which will never be forgotten.

> Not as tragic a country as Romania, but Bulgaria still has enough problems to deter all but the most determined foreign farmer from investing. I returned a few years ago to find Ivan Panayatov still breeding wheat at the Institute for Wheat and Sunflowers. Not much has changed but Ivan was less ebullient than he had been. The Bulgarian countryside has not flourished under capitalism. I wonder if it ever will.

TO THE ENDS OF THE EARTH

THE LAST FARMER IN THE WORLD
1989

'The weather forecast this weekend was so good I recorded it on the video,' said Marit Dyrhaug. 'I'll watch it during the winter to cheer myself up.' A tousled blonde who works as an agricultural Extension Officer in the northern Norwegian town of Sandnessjoen, Marit had arrived in a cloud of dust and a battered yellow Lada. She rolled a cigarette and thought carefully about my question. Who was the last arable farmer in Norway? Back in Oslo the Ministry of Agriculture had suggested that the limit of cereal growing was probably some 60 miles below the Arctic Circle on one of the islands which litter the Norwegian coast. At the same latitude in Siberia, Alaska or Canada farming is impossible but, thanks to the Gulf Stream, grass grows right up to the North Cape of Norway.

The Lada bumped and rattled over Sandnessjoen's potholes as we headed south. Just past the cooperative dairy on the edge of town, Marit pointed out a small, grey house. 'That's your man,' she announced. 'I've checked with my colleagues and nobody grows barley further north than here. I'm certain of it.'

Karl Karlsen was naked from the waist up when he came to the door. The smell which surrounded him suggested that he had fed a lot of pigs and was about to take a shower. A stocky 45 year old, he spoke English with some difficulty but, like all farmers, he was keen to show off his animals and machinery.

Unusually for Norway, Karlsen does not keep cows. 'I used to have 12 of them but I found that milking twice a day was a lot of hard work for very little money so I sold the herd and bought some pigs instead. Life is a lot easier than it used to be.' He grinned. 'Mind you, it's not too comfy today. I've some machinery which I use on building sites. When the North Sea oil business was booming I did pretty well, but this town's gone very quiet now and there's not a lot of work.'

I asked him whether it was true that he grew barley. 'Yes, but not much. I've 45 dekars (one dekar = 1/4 acre), so even by Norwegian standards I'm very small.'

Outside in the farmyard a collection of rusty machinery contrasted with two very new barns. 'We had a fire a few years ago,' explained Karlsen, 'and these all came from the insurance company.' He pulled back one of the sliding doors to reveal a small Claas combine. The tyres were flat and the paint peeling. 'I'm afraid it's not very impressive,' said Karlsen. 'I bought it second hand back in 1975 and it's seen better days. But it doesn't have much to do each year so it'll certainly see me out.'

I tried to conceal my excitement. 'Has it ever occurred to you,' I asked, 'that this is the last combine in the world? Because I reckon that you're the last arable farmer in the world.' Karl Karlsen paused to digest this information. There was no reaction and I assumed that he had not understood my pompous pronouncement.

After more silence he announced gravely, 'It is possible that you are correct.' Had he heard of any farmer who grew barley north of him? 'No. I have not. It would certainly be impossible on the mainland because the weather is much fiercer there. It would have to be on one of these islands.' Did this mean that he was the most northerly arable farmer in Norway? 'Yes, I suppose it does.'

I persuaded him to drive his combine to the nearby field of barley. Why did Karl Karlsen grow this crop when none of his neighbours felt it wise? 'It's probably because I'm interested. You could even say it was a hobby because my harvests are often very small.' He stooped to pick a very green ear of corn which had just emerged from the stem. 'The way things look this year it's possible that I won't have a harvest at all. The summer in this part of Norway has been so cold and wet the crop is at least three weeks later than normal.' He went on to explain that the growing season is so short he can only plant the seed in May and must harvest the barley by late September. 'If it doesn't manage to ripen by then the frosts will kill it completely.'

I wondered how he managed to survive in such circumstances. 'If it weren't for the pigs, I wouldn't,' he admitted cheerfully. 'I sell around 500 a year to the local butchers and the cheques they send me keep my family going.'

The pigs also contributed to the barley crop. Their slurry is the only form of fertiliser Karlsen uses. 'I never have to worry about pesticides either,' he told me. 'It's not that I'm an organic farmer, but simply that this far north we don't get any diseases. In fact we sometimes don't even have a crop at all.' I assumed this was an exaggeration typical of all farmers who enjoy complaining.

Did this happen often, I asked? He stared at the horizon for a moment before choosing his words carefully. 'No, hardly ever, but last year was the exception. I didn't actually have a harvest at all last year. First there were very strong winds and then came the rain. But in normal years I very rarely have a complete crop failure. Of course the yields do vary a lot. The best harvest I've ever known gave me rather more than a tonne per acre but there have been some years when I only got a couple of hundredweight. Then it wasn't really worth getting the combine out of the shed.'

A British barley farmer would expect an average of two tons per acre, and anything less than 30 hundredweight would be disastrous. But if Karlsen's yields were poor, at least the price he received would make an Englishman envious. Norway is not a member of the EEC and so is immune from the Common Agricultural Policy. This enables the Norwegian government to pay nearly three times the price for barley that Brussels has decreed for European farmers.

Karl Karlsen was probably the last cereal grower in the world, but he certainly was not the last farmer. To find him would be a more difficult task. My original idea was simply to drive up to the North Cape and then turn round. On the way south, I assumed, the first farmer I came across would automatically be the man I was looking for. But it was not quite as straightforward as I had imagined.

The town of Alta lies 200 miles north of the Arctic Circle. On the shores of the nearby fjord some of the oldest rock paintings in northern Europe were discovered in 1973. Dating from 4000BC, they depict not only scenes of fishing and hunting but – surprisingly – a herd of reindeer penned into a corral. Here on the edge of the arctic, the first beginnings of livestock farming were apparently flourishing some 6,000 years ago.

Today the farming tradition is still strong in Alta; it contains the most northerly agricultural research station in the world and thus seemed a logical place to start my search for the last farmer in Norway. The question of where I should look was, however, received with total incredulity by the scientists, and I left without any suggestions.

On a nearby farm my luck began to change. I came across the local vet who, together with the artificial inseminator, was making a house call. They found my request all too simple. 'No problem,' said the inseminator, with his arm hidden inside a waiting cow, 'I cover the whole of Finmark and I know all the herds in the region. The man you're looking for is called Jenssen. He lives on Bekkarfjord, nearly a day's drive from here.' I thanked him warmly (on behalf of myself and the impregnated animal) and headed north east.

Steinar Jenssen was milking his 11 Norwegian red cows, some of whom were wearing bovine brassieres to support their very large udders. His wife, wearing a red woolly hat, was helping him in the cow stalls which occupied the ground floor of the barn. In the loft above, the fodder was stored so it could be forked down to the cattle through a hole in the floor.

Communication was difficult until out of the nearby yellow farmhouse appeared a young woman who turned out to be Jenssen's daughter. She was a student nurse in Oslo who spoke excellent English.

Steinar Jenssen farms 46 bleak acres. His fields, which line the shores of the fjord below the wooded hillside, were yellow rather than green because he had just finished cutting the silage crop. In these northern latitudes there is only a single grass cut a year, compared to the three and even four which a Devon farmer will get. 'It's been a good crop this year,' he told me as he forked freshly cut grass to his waiting herd. 'We should have enough to see us through the winter.'

After milking was over we retired to the spotlessly clean farmhouse. I had hoped to be offered a glass of Norwegian poteen known as *Hjemmebrent* (literally Home-burnt) but instead a cup of tea was suggested.

Like Karl Karlsen, Steinar Jenssen had never even considered the possibility of being the last farmer in Norway, let alone the world. When I suggested this to him his face cracked into a grin. 'I suppose it's possible,' he admitted. 'I certainly don't know of anyone who milks cows north of me, but perhaps there is someone.' He rubbed his chin and continued. 'Mind you, I don't envy them because it wouldn't be an easy life. Up here we have so little time in the summer. We have to do all our work in three months instead of the six months they have in the south of Norway where the growing season is longer and the winter shorter.'

There were, I discovered, some compensations. His milk, which is collected twice a week by a lorry from the dairy in Laksalv 100 miles away, is worth 60 pence per litre, three times as much as a British farmer receives. And the milk producers of Finmark, unlike their colleagues in the rest of Norway and the EEC, have no limit on the amount they can produce.

It was nearly nine in the evening but the sun was still high in the sky. 'It may seem pleasant for you southerners to see the sun at midnight,' said Jenssen 'but it makes farming difficult. Don't forget that from the middle of November until late January I never see the sun at all. At around lunchtime there's a faint glow in the sky and that's the nearest to daylight we ever get.' Mrs. Jenssen nodded wearily.

Below us on the fjord a small motor boat was moored. 'I do a bit of hunting in my spare time,' said Steinar Jenssen, 'but what I really enjoy is fishing. Not that I've got a lot of time for that sort of thing. I've just bought ten acres of that hill-side from the government and now I'll have to plough it and get the grass seed drilled.' Why, I wondered, did he need more land? 'To keep more cows, of course,' said Steinar. 'It's not easy making a living with only 11 animals and this extra land will enable me to feed at least two more.'

Throughout northern Norway reindeer, herded by wandering Lapps, roam freely by the roadside. Did these animals compete with the cows for grass? 'Not normally,' said Steinar, 'but there are times in the early spring when there's still snow on the mountains and the reindeer come looking for grass lower down. But I'm lucky because on this farm I'm in between two groups of Lapps and there aren't too many reindeer round here. In other parts of Finmark dairy farmers have been known to shoot reindeer when they find them eating the grass. Then there can be serious trouble both with the Lapps and with the police.'

What did he do all winter, I wondered? 'Oh I keep myself busy, don't worry about that,' came the answer. 'There are a lot of repairs to do to the buildings and the machinery, and don't forget that the cows still need to be milked twice every day of the year.' Was he lonely? 'No, not at all. I was born here and so this climate is natural for me. In fact I'd find it difficult in your part of the world. Too many people, too much noise. Besides,' he added, 'I've got company up here. There are three other farmers on this fjord and we've a lot in common.'

Mrs. Jenssen, who had sat quietly throughout the discussion, was becoming restless. She whispered a few words to her husband. He looked up and announced, 'It's time to feed the calves.'

As I left, Steinar Jenssen was getting into his overalls and his wife had again put on her red woollen cap. I turned to look back across the fjord. There was still snow on the distant mountains and the wind had whipped the water into whitecaps. In the distance I could just make out a small figure walking towards the barn with a bucket in each hand. Steinar Jenssen certainly looked like the last farmer in the world.

> This journey was the beginning of my obsession with the Arctic – from which I have never recovered. There is something eerie and magical about parts of the world where even farming comes to a halt.

GREENPEACE vs GREENLAND
1996

As the Arctic tern flies Qaanaq is 800 miles from the North Pole, which makes it the last and most northerly settlement on earth. It is located on the north west coast of Greenland separated by 50 miles of ice from Ellesmere Island in Canada.

Three thousand years ago the first inhabitants of northern Greenland, whose ancestors had walked across from Siberia, moved into this region. Then, as now, they hunted polar bears, Beluga whales, narwhal, walrus and seals. On land they trapped the polar fox and arctic hares. The skins were used for clothes, the meat for food and the bones for implements. The Greenland Inuit remained in this time warp until, sometime in the sixteenth century, they first came across European sailors looking for whales. The white men brought civilisation in the shape of metal knives, tobacco, syphilis and – most lethal of all – alcohol.

Qaanaq today is a small cluster of pale blue and yellow houses with steep Danish roofs and neat front doors. The village lies scattered on the northern shore of Murchison Sound. Opposite is the Politiken Glacier (named after the Copenhagen newspaper) and behind it the endless Greenland Icecap which is up to two miles thick. If this icecap ever melted the sea level round the world would rise by 25 feet.

The main street of Qaanaq consists of two shops and the central administration building. Down on the shore racks of needle-sharp kayaks remind one of the boathouses at Putney. On the hillside each house has its own water supply outside the front door. A chunk of ice, sometimes ten feet tall, stands in every yard. From these miniature icebergs are chipped slivers which are then melted in the kitchen. Behind the houses where a garden would stand in Denmark is a small wire mesh compound where the household's huskies live, howling at night and sleeping during the day.

For three winter months Qaanaq lies in total darkness with temperatures which fall to minus 40 degrees Celsius. In summer the snow disappears from the shore and the sea ice melts, leaving only the occasional iceberg to float southwards into the sun.

Although the village now has electricity, satellite communications and even a tiny supermarket, it is unique among arctic communities. Qaanaq is the last settlement in which the Inuit still hunt just as their ancestors did thousands of years ago. Unlike their cousins in Canada and Alaska, the hunters of Qaanaq still use kayaks not outboard motors. Instead of noisy ski-mobiles they use the traditional dog teams. This makes Qaanaq the only working example of traditional Inuit culture left in the Arctic today. It also makes Qaanaq one of the most precious and most fragile communities on earth

But today Qaanaq is dying and the culture is dying with it. The occasional old hunter does still get into his kayak and paddles down the fjord looking for walrus and narwhal. A few other old men still harness their dogs to the sledge and drive out onto the ice edge in search of seal or polar bear. But back in Qaanaq the young men sit at home. Out of a population of nearly 500 there are no more than five young hunters left in town today.

The reason for this decline is not the advent of television or the effects of a Danish education, it is because Greenpeace, the self-appointed guardian of planet earth, needed cash. Lots of cash. It all began more than a decade ago in Newfoundland (some 2,500 miles from Qaanaq) where the local fishermen decided to kill the seals which had been eating their catch. Photographs of round-eyed baby seals lying in pools of red blood on the white snow flashed round the globe and the civilised middle-class world was horrified and sickened.

Greenpeace, which knew perfectly well that seals were not a threatened species, immediately recognised the potential of the photographs. The image was powerful. Big greedy humans battering defenceless baby seals to death just to protect their income. Within months Greenpeace had organised a campaign to stop the trade in sealskins. It was a brilliant success and culminated in the European Parliament banning all imports of sealskin into the EEC. Meanwhile, of course, there was another effect which for once Greenpeace did not publicise. Cash poured in from people who wanted to stop big-eyed baby seals being battered to death.

In the days before civilisation reached Qaanaq the Inuit did not need much money; they were almost totally self-sufficient. Today, however, they live in a cash economy, which means they have to sell something to produce an income. And the only crop the Qaanaq Inuit can harvest in any quantity is the seal. As a result of Greenpeace, the price of a good sealskin has dropped from £70 to £3, to which the Greenland government today adds a £22 subsidy. Meanwhile the cost of living has continued to rise.

'We call them Greenshit,' said a hunter's wife I met as she queued for her family's social security payment in Qaanaq's Government Building. I had gone there to visit one of the town's most famous inhabitants, an Inuit called Robert Peary, the great grandson of Admiral Robert Peary who passed through Qaanaq on his way to the North Pole in 1909 and apparently had time for some rest and recreation en route. 'I know Greenpeace does some good in the world,' Peary told me, 'and I know it protects the animals. But perhaps they should think about human beings too.'

Bo Norreselet, a Danish administrator, was less charitable. 'Greenpeace just lives in a different world to us, but even when they came here to Greenland they still didn't understand.' He explained that at the height of the anti-sealing campaign Greenpeace representatives had been invited to a public meeting in the small settlement of Ummannaq. 'The afternoon Greenpeace arrived the village hunters returned from a trip and started unloading the seals and narwhal on to the beach. Greenpeace assumed this was a publicity stunt for their benefit and refused to believe that it was a normal occurrence in any Inuit village.' At the meeting later that day they were asked what the villagers should do instead of catching seals. The Greenpeace spokesman suggested that they grow crops, and only when he was taken outside and asked to dig into the gravel did he admit that this was not a very good idea.

I asked Norreselet how Qaanaq had reacted to Greenpeace. 'We were very angry and very confused,' he said. 'We tried to explain our situation in Europe because we felt that the whole truth had not been told. But we were amateurs up against Greenpeace. They knew how to handle the press and the television. They just kept putting out pictures of baby seals with blood on the snow and nobody wanted to hear from us. We tried to explain that here in Qaanaq we didn't club baby seals to death. There aren't any baby harp seals here anyway. But again, nobody would listen to us.'

What, I wondered, would Norreselet like to tell Greenpeace today. He paused, rubbed his chin and stared out across the frozen fjord. 'I would tell them' he said slowly, 'to stop threatening the life of this society. Don't measure all other cultures by European standards – using your morals to judge societies which are not quite as developed as your own. Accept Greenland's struggle to live and the concern Greenland has for its own animals.'

Kaj Soby, the local schoolteacher, continued my education. 'The seal to us is like a pig or a sheep to a European peasant. We use every part of the animal, either as food or as clothing just like some Europeans use every part of a pig. We have been doing this for nearly 3,000 years in this part of Greenland and our lives have changed very little. The seal is part of the Greenland culture. We use it in so many ways you find it impossible to understand when someone from Greenpeace tries to explain why we should no longer hunt seals.'

By then Soby was growing bitter. 'There's a double standard here. In Europe you kill cattle and use the skins for your shoes and your coats but now you tell us that we can no longer kill seals. You may be able to kill your cattle in a more humane way than we kill our seals. You can use the skin of calves to make leather upholstery but today we are not allowed to sell the skin of seals for the same purposes – unless we remove all the fur and sell the result as skin. That's another double standard which you rich Europeans have imposed on us. You won't buy our sealskin unless we first remove the fur and then – as if by magic – it becomes perfectly acceptable.'

'The problem is that Greenpeace has power. It has the power to tell the western world what is right and what is wrong. And most people living in Copenhagen and London believe them. How can we manage to compete against them and tell our side of the story? Nobody wants to hear what we have to say. Nobody wants to look at a picture of an Inuit family trying to stay alive but everyone wants to look at a picture of a baby harp seal bleeding on the snow.'

How badly hurt is Qaanaq today? I asked Soby. 'If you remove the possibility of a society to live by what it produces, you turn people into beggars. As a result of Greenpeace's campaign all those years ago we are now a village of beggars. We ask for support from the Greenland government, from Denmark and from the EU. Maybe we should ask for support from Greenpeace. Today our pride and self-esteem has been taken from us. Nobody needs our skills any more so our only real job is to go to the government office to collect our social security.' There was a pause. 'Or wait for the monthly alcohol ration to come in. We're pretty good at drinking up here.'

This was the first time I had ever encountered the ugly side of Greenpeace. Today, as I watch them rip up farmers' crops, I am no longer surprised. At the time I thought of Greenpeace as being an environmental Oxfam. Now I know better.

MY SEARCH FOR MANDELSTAM
1993

All I can do, therefore, is to gather what meagre evidence there is and speculate about the date of his (Osip Mandelstam's) death. As I constantly tell myself: the sooner he died the better. There is nothing worse than a slow death. I hate to think that at the moment when my mind was set at rest on being told at the post office that he was dead, he may actually have been on his way to Kolyma. The date of death has not been established. And it is beyond my power to do anything more to establish it.

(The last paragraph of Nadezhda Mandelstam's *Hope Against Hope*)

The Chairman of the Magadan City Soviet was restless. He looked at his watch and fiddled with the lock on his briefcase. I had been making small talk about the weather (minus 30 degrees Celsius) and his job (due to be abolished by Yeltsin's new Constitution) but the time had come to get down to business. 'What about the Mandelstam papers?' I enquired in a voice which I hoped did not betray my anxiety. There was a pause while Leonid Musin looked out of the window and scratched the back of his neck. 'I don't have them any more,' he replied. 'They're back in the archives.' Perhaps he could retrieve them? The answer was bleak. 'I'm not sure where they are. Maybe they've gone to Moscow.'

Two months earlier I had been sitting in Musin's office on the top floor of Magadan's City Hall. On the long conference table was a map of the goldmines along the Kolyma river and after ten minutes of statistics my mind was becoming numb. As the translator droned on I studied the picture on the wall behind the Chairman's desk. In the old days a benign Lenin would have stared bravely into the future. Now he had been replaced by a colour photograph of Anchorage, Alaska by night.

Suddenly, and for no apparent reason, Musin rose from the table, walked to a small anteroom and returned with a pale brown folder which he placed delicately onto the table in front of me. I assumed it contained some historical records of gold production or municipal housing in Magadan so was in no hurry to examine it. Sensing my reluctance, Musin opened the folder. Inside was a faded mug shot of a balding man with tired eyes and a truculent chin. On the facing page was a form completed in blue ink with a solitary fingerprint. Beneath the photograph, written in white, was the name Mandelstam. O. I was looking at the Gulag file of Osip Mandelstam.

Inside was a bundle of flimsy forms, some typed and some covered with notations in red crayon. Two more pages of fingerprints and then a handwritten letter. It was from Nadezhda. Could I make a photocopy? 'Nyet.' Could I take a photograph? A long pause and then a hesitant 'Da'. But only the front two pages of the file. Nothing more. What, I wondered, did Leonid Musin want from me? Another pause. 'We are looking,' he said slowly, 'for sponsorship from the west. Sponsorship to publish these documents from the archives and sponsorship to erect a memorial to the victims of the Gulag here in Magadan.'

Back in England I told the story to every Russian expert I could find. Their verdict was unanimous: Leonid Musin had been speaking in code. Like so many Russian functionaries these days, his real objective was cash. Had I offered dollars I would have been allowed to copy the file. Off went a fax to Magadan City Hall. Could I return and take a further look at the Mandelstam papers? The reply was immediate. Yes, the file was still with the Chairman and he would be happy to see me.

Magadan was built by Stalin in 1932 with only one purpose; to administer the coldest, remotest and harshest part of the Gulag Archipelago. The region, named after the river which flows into the arctic ocean, is called Kolyma. Even today, 40 years after the Gulag was closed, the name Kolyma makes most Russians shudder. Like Auschwitz, it has come to represent not just a place but a whole system of inconceivable brutality. Solzhenitsyn, who almost omitted Kolyma from *The Gulag Archipelago* 'because it was a whole separate continent,' called it 'the pole of cold and cruelty'.

And nowhere was crueller than Serpantinka. Every other concentration camp in Kolyma had an economic purpose; to mine gold, tin or uranium. But Serpantinka had a different function. It was the slaughterhouse of the Gulag. Named because of the road which snaked down from a low plateau towards a gold-rich river valley, it consisted of only a few sheds surrounded by barbed wire and wooden watchtowers. Beside the sheds a small ravine ran steeply down towards the camp of Kattenakh some two miles away.

I had travelled 400 miles north from Magadan to the small town of Elgen which, like every settlement in the area, had once been a concentration camp. The temperature in early November had fallen to minus 41 degrees Celsius and the aspen trees were leafless. Low hills separated wide valleys in which the spoil from the gold dredgers had blocked the rivers.

My jeep stopped on a hairpin bend above which a small ridge stuck out into the valley. Twenty feet above me was a snow-covered chunk of grey rock. As I climbed nearer I saw that the stone had been wreathed in barbed wire and at its foot were some plastic flowers and an empty vodka bottle. I cleared away the snow and found a black marble slab on which had been chiselled the word SERPANTINKA.

Back in Magadan I stayed in one of my favourite hotels in the world, the romantically named Business Centre. Three years earlier it had been the Communist Party's VIP Hostel. Today, run by the former boss of the Magadan Konsomol, it has 17 rooms, a small dining room and a cosiness which is rare in the rest of the world and unique in Russia.

But that afternoon I was feeling anything but cosy. I had flown half way round the world in response to Leonid Musin's fax and now I found the Mandelstam file had disappeared. My Russian experts had been wrong; the dollar bills I carried in a pouch round my neck remained untouched. As Musin left the hotel, promising to make some phone calls, I knew I wouldn't hear from him again.

Natasha Sokolova, with a cardigan slung over her shoulders and a glass of tea in her hand, listened to my story, paused and smiled the sort of sympathetic smile that only good schoolteachers can produce. 'We've got some children here at School Number 17 whose parents work for the KGB and I will try to contact them for you,' she said. 'I suppose it's just possible that they will let you into the Gulag archives and since I've never been there myself I'd be happy to come along too.'

The following morning Natasha and I stood outside a pale yellow two-storey building with bars on the windows. After three knocks the door was opened by a man whose grey hair matched his suit. 'Welcome,' he said in a whisper. 'How can I help you?' Gennady Petrovich Korneev was the Director of the Ministry of the Interior Archives who, with a staff of 12, worked in what had once been the central prison in Magadan. He led me down steep steps to the basement where a row of cells lined a dimly lit corridor painted institutional blue and white. Along the walls, stretching from floor to ceiling, were wooden shelves divided into compartments. In each compartment lay a neat heap of the pale brown folders. I was in a morgue. Each folder had once been a human being.

There was a tap on my shoulder and I turned to see a dark-haired woman with vermilion lipstick. 'I think this is what you are looking for,' she said, handing me the folder I had last seen on Leonid Musin's conference table. On the front cover written in black ink were the words File No. 117794. TO BE KEPT FOR ALL TIME. Natasha leaned over my shoulder and translated as I leafed through the documents. Among them was the following letter.

7/2/39

To the main administration of camps

I have learnt that my husband, convict Mandelstam Osip Emilievich died in Vladivostok SVITL, Barracks No. 11 (five years for counter-revolutionary activity), since money I sent has been returned to me 'because the addressee is dead'. The date of death is being given as being sometime between 15/12/38 and 10/1/39.

I ask the administration of the camps to check whether this information is correct and to issue me with an official certificate concerning the death of O.E. Mandelstam.

(signed) Nadezhda Mandelstam

I request that the reply be sent to the following address:
Moscow, Starosadsky 10, Flat 3
Alexander Emilievich Mandelstam

At the present time I have no address, since my temporary residence permit for Moscow has expired and I am looking for accommodation near Moscow.
A few pages later I found a small thin piece of paper with two signatures. It was the Death Certificate:

Arrived from Moscow: 12/10/1938
Placed in the camp hospital: 26/12/1938

We the undersigned Doctor Kresanov, Duty Medical Attendant have compiled the following document concerning the death in hospital of the separate camp centre of the SVITL NKVD.

Surname and patronymic: MONDELSTAM (sic) Osip Emilievich
Year of Birth: 1891
Place of birth: Poland
By whom and when convicted: Special Session of the NKVD USSR 2/VIII 38.
Article and length of sentence: CRA (Counter-revolutionary activity) 5y.
Most recent place of residence: Kalinin
Cause of death: Paralysis of the heart and arteriosclerosis. The corpse's finger-prints were taken 27/XII-38. In view of the fact that the cause of death was clear, the corpse was not subjected to post-mortem examination.

Signed: Kresanov
Signed: (Illegible)

    Reading this document I realised that Nadezhda's story was not quite as straightforward as it once seemed. For some strange reason the NKVD did exactly what she had requested. They sent the death certificate to Osip's brother, Alexander, who was summoned to the Registry Office of the Bauman District in Moscow to receive it. By then, however, at least four accounts of Mandelstam's death were circulating and Nadezhda, confused by so many different myths and profoundly suspicious of the authorities, continued to search for the truth which she already knew.
    Gennady Korneev had been waiting quietly, like an undertaker watching while the last respects were paid. I closed the file and returned it to him. None of us said anything. Outside in what had once been the prison yard the snow was beginning to fall. From far below in the basement came the sound of a heavy door closing. Magadan is full of echoes.

    I had gone to Magadan simply to pass the time while I waited in Seattle for some friends to arrive. I noticed in the newspaper that Alaska Airlines had just started a direct flight to Magadan and it seemed like a good place to spend a few days. Little did I imagine that I would discover the fate of one of the great literary figures of the twentieth century.

THE NORTHEAST PASSAGE
1991

At precisely midday on 13 June Sergei Sidorenko, chief mate on the nuclear ice-breaker, *Arktika*, pushed three throttles forward. The ship, with its black hull and orange superstructure, moved slowly away from the dockside at Murmansk's Atom Base. Its destination was somewhere north of the Bering Straits where it was due to meet a convoy of cargo ships and escort them through the ice to their destinations on the arctic coast of Siberia.

My objective was even more distant: the port of Vladivostok, some 6,000 miles and eight seas away. I was excited not simply because of the distance, but because the last time a foreigner had been permitted to make this voyage was during the brief Hitler–Stalin pact when, in 1940, the German-armed raider Komet was helped by Russian icebreakers to sail from Norway to the Bering Straits, from where it slipped unnoticed into the Pacific. Since then the Soviet Arctic has been a closed military area.

For a day and a half the sea looked more like the Caribbean than the Barents, flat calm under a blue sky. Only the near-freezing temperature reminded me that we were 200 miles north of the Arctic Circle and at the furthest reaches of the Gulf Stream. On the second night the captain, Gregorii Ulitin, threw a party to celebrate my 50th birthday. Any fears I might have had that *Arktika* was a dry ship evaporated. Under a benign Lenin wearing a blue polka-dot tie, we drank to me, the Queen, the Russian merchant marine and anybody else we could think of, including Boris Yeltsin, who had just received the votes of 80 per cent of *Arktika*'s crew. The food was good and the vodka exceptional since it was the Siberian brand which is 5 per cent stronger than the normal tipple.

The first signs of ice looked, from a distance, like dandruff on the sea. Small ice-floes, no more than three yards across, grew more frequent as we neared the southern tip of Novaya Zemlya. On the larger ones seals were basking in the sunlight of the 24-hour polar day. When these miniature icebergs hit the hull they made a noise like a distant hammer which could be heard throughout the ship.

Behind the bridge was a small, dark room full of electronic equipment and festooned with illustrations of NATO warships. It was referred to as the Chaikana because this is where the tea was brewed, but its original purpose was more sinister. It had been designed as the fire-control centre for the icebreaker which, in case of war, would be converted into an armed cruiser. It was here that I came across the bearded second mate, Maxim Shumilov. He did not share my excitement at this first sign of ice. 'In a few hours,' he said, 'You'll see the real stuff. The diesel icebreaker, *Captain Sorokin*, has just radioed that she's hit pack ice and can't continue with the freighter she is escorting.'

At supper that night in the officers' dining room when *Arktika*'s bow struck the wall of ice, the soup slopped out of the tureen. The narrow straits leading from the Barents to the Kara Sea were completely blocked. A few hours later we met the *Captain Sorokin* and her charge, the orange-hulled freighter, *Captain Danielkin*.

In the following 24 hours *Arktika* managed to travel 42 miles, in contrast to the 350 miles we had covered the previous day. In pack ice, some of which was four metres thick, the two nuclear reactors were producing 90 per cent of their 75,000 horsepower. Yet even then we were often forced to a complete standstill. When this happened the ship would go astern for 300 metres before hurling herself at the ice again. The specially shaped bow caused us to ride up onto the surface of the ice before the downward motion of the ship cut through to the water.

The sensation of breaking thick ice is not a pleasant one; it is reminiscent of riding a tractor across a ploughed field. Unlike the movement of other ships, those of an icebreaker are irregular, unpredictable and extremely uncomfortable.

Three days later we reached the port of Dikson at the mouth of the Yenesei where the Captains *Sorokin* and *Danielkin* bade us farewell. That evening, as we turned northeast along the coast of the Taymyr Peninsula, I spotted a polar bear with two cubs wandering along a pressure ridge of ice. In the days to come I would see many more bears, but none was quite as exciting as the first I saw in the Arctic.

The ice grew thicker and more unyielding. Gradually I found I no longer noticed the sudden movements of the ship, and did not wake throughout the night every time we hit a particularly hard patch. Two or three times a day Valerii Losef, the most experienced ice-navigator in the Arctic today, would take off in *Arktika*'s helicopter and would fly ahead of the ship trying to find any cracks or signs of weakness through which to steer.

I accompanied him on many of these flights in what the crew called their *Ptichka* (little bird) and sat fascinated as this man studied an apparently unbroken desert of ice, noticing features which I could not even see when he pointed them out to me. Occasionally we would land on remote islands to visit the graves of long-forgotten Arctic explorers or to sift through the remains of expeditions' stores at campsites which had been deserted for nearly a century. On other occasions we would land in sheltered valleys where, in late June, small Arctic flowers had blossomed, lichens were orange, yellow and red and birds had laid eggs in softly camouflaged nests.

It was at these moments that I would begin to understand what makes the Arctic so magical. In these places 1,000 miles from the nearest town, with the air cold, clean and (when there is no fog) excruciatingly clear, there is a quality of silence I had never heard before. Our stops were always too short.

On our way back to *Arktika* after a morning's work in the helicopter, we would sometimes disturb an unsuspecting reindeer, swoop down low to look at a polar bear, or hover as a seal lay like a huge slug on the surface below us.

After a week of solid ice we eventually reached Cape Chelyuskin, the most northerly point on the Eurasian land mass, a mere 780 miles from the North Pole. Strangely, as we crossed from the Kara to the Laptev Sea, the ice vanished as suddenly as it had appeared. Once again we were back in partially open water dotted with ice-floes. On some of these ice-floes colonies of walrus lay back and watched while *Arktika* came within yards.

By now we had received new instructions from Murmansk. We were to head for the port of Pevek where we would find a convoy waiting for us. The ice pilot's reconnaissance showed that the best route lay to the north of the New Siberian Islands so, instead of hugging the coast, we steered due east.

On 21 June, the longest day of the year, I wondered how high the sun would be above the horizon. The captain, a gruff veteran of 30 years in the arctic, who could also quote Burns and Byron in Russian, told me the answer. 'Your fist (*Kulak*, in Russian) is eight degrees, and the outstretched fingers of your hand are sixteen.' The sun appeared to be one-and-a-half kulaks above the horizon. A quick check with the sextant confirmed my measurement: 12 degrees.

I was, as it turned out, lucky to see the sun because for the previous four days, and for most of the next week, a cold, damp Arctic fog settled over the ice, lowering both visibility and my morale.

In the East Siberian Sea we found ourselves back in heavy ice, but this time it was unlike anything we had seen before. Instead of being white, or pale turquoise blue, this was dirty brown stuff covered with black puddles. It was, I learned, two-year-old Canadian pack ice which had drifted across the north pole and was now moving south towards Siberia.

Nearly three weeks after leaving Murmansk we reached the small Siberian town of Pevek. Founded as an outpost of the Gulag, it is today the most important port in northeastern Siberia.

After a day's rest we headed westwards, escorting a small coaster with a cargo of timber to the mouth of the Kolyma River before turning back once again to Pevek. The ice pilot returned that afternoon with the rarest of prizes – fresh fish. He had landed at the mouth of the Kolyma where he had met a native Chukchi fisherman. That evening we held a secret party in the Chaikana. The menu consisted of vodka and raw Ryapushka, a bony but delicate freshwater fish the size of a herring.

By now I was becoming restless. My original plan had been to find a ship in Pevek which was heading east to Vladivostok, but a problem had arisen. Because of a bureaucratic mistake, my visa did not mention Pevek and I was not permitted to go ashore. It was beginning to look as if I would have to stay aboard *Arktika* until she returned to Murmansk in four months' time.

Just when I was beginning to resign myself to this fate, we were ordered to return to Pevek and escort a tanker to the Bering Straits. A few more radio exchanges and eventually a slightly bemused captain of the tanker, *Igrim*, agreed to accept an itinerant Englishman and deliver him to Vladivostok.

My actual emigration took place three days later in the Chukchi Sea, south of Wrangel Island. In heavy ice *Igrim* halted and *Arktika* went astern so that the tanker's bows nuzzled the icebreaker's stern. After some tears and much kissing, I clambered up a ladder on to *Igrim* and waved goodbye to *Arktika* and her crew.

Two days later I woke to see the most easterly point of the Soviet Union, Cape Dezhnev, off the starboard side. Partly hidden in the fog on the port side was Great Diomede Island, and behind it the United States in the shape of Little Diomede Island.

As we sailed south through the Bering Sea, sights which I had once taken for granted began to reappear. One evening I became aware of twilight, and a day later the sun dipped below the horizon.

The Sea of Okhotsk was invisible. Fog, thicker than anything I had seen in the Arctic, seemed to cover the world. Only after passing the La Perouse Straits between Sakhalin and Japan did the sun reappear. By then *Igrim* was in the Sea of Japan heading for home as we skirted the supremely beautiful Siberian coast with its green hills looking like the Mountains of Mourne.

Six-and-a-half thousand miles after leaving Murmansk, I found myself along-side the oil jetty in Vladivostok. I had been the first Briton ever to transit the northeast passage, and the first foreigner for over half a century. Perhaps I should not have been surprised when the receptionist at my hotel in Moscow grew irritated by my explanation of why I had no stamp in my visa since arriving in Murmansk almost six weeks earlier. 'It is,' she announced, after consulting with her colleagues, 'impossible to travel from Murmansk to Vladivostok by ship.'

Only when I returned to England did I learn (to my amazement) that I had been the first Englishman ever to go through the north-east passage. The friends I made on *Arktika* are some of the finest men it is possible to imagine. Which is why I was so shattered to hear that Valerii Losef had been killed in a helicopter accident while flying his *Ptichka* in March 1999. He was both a friend and a hero.

POSTSCRIPT

CATHERINE WALSTON
1994

Every little boy thinks his mother is the most beautiful woman in the world and I was no exception. The difference is that even today when, as a middle-aged man, I look at the photographs of her which have been recently published, I have no reason to change my mind. And this is in spite of the fact that for the last decade of her life she was a very sick lady indeed. The effort and excitement of burning her candle not just at both ends but also in the middle must have taken a terrible toll.

The strange thing is that even in her final days when my mother's mind wandered and her body was no more than a shell, the men in her life remained constant and devoted. Chief among these was, of course, my father. They had been married 42 years earlier and – in spite of the sort of pressures which would have smashed most marriages – they shared a family, a home and a bedroom.

Another man also seemed still to care about her. Graham Greene lived 800 miles away in a small apartment overlooking the port of Antibes. At irregular intervals letters used to arrive addressed to my mother written in that strange spidery writing which was, like a fingerprint, unique to Greene.

I had first met him as a small boy of seven when, in the spring of 1948, my mother told me that we were going to Italy to stay with Graham. Together with my younger brother and sister and Twinkle, our nanny, we flew off from Northolt to Naples. For three months we sat in the sunshine of Capri, playing in the walled garden of Greene's white-painted villa. In the mornings we were confined to the furthest corner of the garden and told to be quiet because 'Graham works in the morning and he doesn't like any noise.' In the evenings we would stroll down to the piazza of Anacapri and eat a dish which, for a boy who lived in dreary, rationed postwar England, was unspeakably exotic. It was called a pizza. On rare occasions we would go on expeditions, sometimes by boat to the Blue Grotto and sometimes to the other side of the island to visit Gracie Fields and her husband who had settled there.

Greene himself was a distant figure who appeared to tolerate children but never to enjoy us. He must have looked on us as a price he had to pay to have my mother's company. And this in itself was strange since my mother, for all her energy, excitement, generosity and unpredictability, was never a good mother in the accepted sense of the word. She liked her children in short sharp bursts between the exciting episodes in her life. Twinkle, in whom my mother used to confide at great length, did all the things which most mothers take for granted; bedtime reading, sewing on Cash's name tapes, listening to our fears and triumphs. My mother would appear, preceded by a bow wave of Guerlain's Mitsouko perfume, and enthral us with an account of where she had been or was about to go.

Throughout our stay on Capri it never occurred to me that Greene was anything more than a good friend of my mother's. The complexities of human relationships, still more the concept of adultery, meant nothing to me.

As Easter approached we left Capri and drove to Rome where – as usual with my mother – we stayed at the best hotel in town, the Hassler at the top of the Spanish Steps. Those were the days of currency restrictions and so we were unable to afford any meals in the hotel. Every morning Greene would lead a procession of three adults and three children down the Spanish Steps to a coffee house where I would have a cup of hot chocolate topped with an unimaginable amount of whipped cream. On the way back we would stop to gaze into the windows of the Perugina chocolate shop and marvel at big chocolate Easter eggs which back home in England would have used up all our sweet rations for a year or more.

In the years which followed, Greene's affair with my mother settled down into something like comfy normality. I saw no signs of the tension between my parents to which Greene's letters refer. To me he was just one of a coterie of friends who came down to Thriplow most weekends to get away from London, sit in the sunshine, read the papers and drink a bit. Whatever passed between my mother and father did so behind closed doors and not a ripple nor an echo ever penetrated the nursery.

During this period my mother travelled the world with Greene, going to the Caribbean to visit Noel Coward, to Vietnam where they smoked opium and, most frequently, to a small cottage on Achill Island off the west coast or Ireland. My father had by then come to terms with Greene and, although they never had a warm relationship, at least tolerated his presence at Thriplow. By then my father was making a gradual transition from farming to politics, spending more time in London where he had a small house above Locks the Hatters in St. James's Street. He too had his fair share of women friends with whom he used to wander round France, staying in simple hotels and eating at expensive restaurants.

After their travels my parents would return to Thriplow like homing pigeons after a mission. There they found contentment and a quietness which must have contrasted starkly with the non-stop action of their lives away from home. The letters between them at this period show an eerily friendly relationship which seemed to have no secrets and created no jealousy. My mother would write long letters from Italy in which she would describe the weather on Capri, how many words Graham had written that day and whether or not he was passing through one of the fits of melancholia which punctuated his life like storms in the Bay of Biscay. Her descriptions were graphic but everyday, almost as if she was describing a day's outing to Brighton with an old friend. My father would reply with news of the farm and the family and would end by saying how much he was looking forward to collecting her from Heathrow the following week.

As I grew older I began to look on Greene as more than just a weekend visitor. One day at my prep school I was leafing through *Picture Post* when my mother's face jumped out of the pages. There she was in her inevitable mink coat standing beside Graham at some theatrical event in Brighton with Noel Coward. I felt half embarrassed and half proud that my mum was famous – or at least had made it to *Picture Post* which came to the same thing.

By then Greene had bought the flat next door to ours in St. James's Street, and when we moved from there to the Albany, he moved too. Very occasionally when Greene was abroad and our apartment in Albany was too full, my mother would use Graham's flat as overflow accommodation and I was given a sleeping bag and told to doss down on the floor. Those were exciting nights. I would count the miniature whisky bottles on the mantelpiece and, after drawing the curtains, would put Eartha Kitt on the record player and read Greene's copy of *Fanny Hill*, which my mother had told me was pornographic. And perhaps it was, but to a pubescent like me Eartha Kitt had an infinitely more disturbing effect.

The last trip I made with Greene was in the summer of 1956. The previous winter, while skiing in Gstaad with my family, I had read in the continental edition of the *Daily Mail* that Grace Kelly was to be married to Prince Rainier of Monaco. For the past two years I had suffered from an acute adolescent crush on the blonde actress from Philadelphia, and went running to my mother to break the dreadful news. Any normal mother would have smiled benignly and asked me if I had done my homework. My mother was not such a person. She immediately booked a call to England and, a few hours later, was speaking to Mrs. Young who at the time was a secretary shared by my father and Greene. Mrs. Young was instructed to book rooms at the Hotel de Paris in Monte Carlo and to make the necessary plane reservations.

Six months later Greene, my mother and I flew to Paris where, after a lunch of *truite aux amandes* at the Restaurant Voltaire on the Quai Voltaire, we caught the train for Monte Carlo. We stood among the crowds which lined the procession route and saw the open Rolls Royce take the couple to and from the cathedral. My agony was heightened by the discovery that Grace Kelly had brought her pet poodle from America to sleep at the foot of the marital bed. The dog was called Oliver. It was more than I could bear.

By then I was an acne-infested Etonian who should have been wise to the ways of the world. Maybe I should have noticed that Greene and my mother shared a room. Maybe I should have disapproved. Maybe I should have sniggered. I did none of these things. I was not only innocent but I had also grown used to Greene's presence over the past decade. He might not have been part of the furniture but he certainly was part of the landscape.

Yet even the landscape changes, and gradually Graham Greene faded from our lives. I would like to pretend that I noticed a great change in my mother. But at the time I was far too self-obsessed with my adolescent traumas to detect the seismic shock waves which result when an old and profound relationship breaks up. Looking back, and using the hindsight which comes to us all if we wait long enough, it is plain that my mother's long and painful decline started at almost exactly the time that she and Greene parted in 1959.

To all outward appearances she must have seemed like the same bewitching, exciting, unpredictable, shocking and sexy woman she had always been. She still travelled widely, entertained lavishly and – most important for her – was still able to turn every male head when she walked into a crowded room.

Her circle of friends had been changing imperceptibly throughout the 1950s and now became more and more centred on an assortment of Catholic priests who used Newton Hall (to where we had moved when the army finally vacated it), as a combination of rest home and restaurant. Most of these were poor tired men whose lives were humdrum, celibate and spartan. Few of them were able to cope with a hostess whose perfume pervaded the house like secular incense and whose scarlet toenails and flashing eyes must have made Jezebel seem domesticated.

But it was not simply my mother's physical presence that all men found so irresistible. Thanks to Greene and his friends, she had read widely and voraciously so that she was as happy talking about the novels of Thomas Mann as the theology of Thomas Aquinas. John Rothenstein, who was then the Director of the Tate Gallery, had continued her education by introducing her to the world of contemporary artists. As a result it was my mother rather than my father who used to visit Henry Moore with brown envelopes full of five pound notes and return with a small bronze and a gouache. Eventually, armed with her own innate good taste and my father's chequebook, she managed to build up an extraordinarily good collection of modern British paintings.

My mother was, in other words, anything but a bimbo. No wonder those tragic priests must have discovered a whole new dimension to temptation for which their spiritual training left them totally unprepared. But in her own way my mother was equally unprepared for life in general and middle age in particular.

One day, coming home from the flat which she kept in Dublin as her very private refuge, she slipped and broke her hip. Within a few months she changed from being hyperactive to a near cripple who walked unsteadily with a stick. Alcohol, which had long been important in her life, now became an obsession. Eventually even the priests drifted away and she sank slowly but inevitably into a half life of pain and depression. Very occasionally, when an old friend or a letter from Greene arrived, she would rouse herself, sit up in bed and for a few hours one could see faint signs of the sparkle which had illuminated so many lives.

Throughout this final tragic period one man stood by her and slept by her. My father, in spite of everything, remained constant in his own way. During her last night I sat in the hospital room while my father stroked her hand and talked quietly about the old days. With tubes sticking through her papery grey skin, she drifted in and out of consciousness. From time to time she would open her eyes, smile an exhausted smile, and close them again. Towards dawn she died and my father sat silent.

Only then was their marriage finally over.

INDEX

Sokolova, Natasha 144
Soviet Arctic 146–9
Soviet Union 132–3
Soyl, GPS-controlled quad bike 25–7
States, Alan 61–4
Stewardship Scheme, Countryside Commission 46–7
Stoyanov, Minko 122
subsidies
    abolished in New Zealand 9–10
    acreage payments 13
    EC 9–10
    effects on production 12
    green 44–5
    and respect from townsfolk 10
sugar beet crowns 37–9
surpluses, becoming shortages 14
Svarstad, Anders 67–8
Szegeds, Hungary 92–7

Todorof, Angel 131
Toma, Nicolae 120
Townes, James 57
Tregar, Lothar 78–9
Twidale, Matt 38–9

Ulitin, Gregory 146–9
USA
    cereals, practices and prices 61–4
    cotton industry 53–9
    fish farming 58–9

Vassily Gaig 111
Vladimirov, Casimir 128–9
Vladivostok 149
Vnuk, Marian 107

Walston, Catherine, *obituary* 151–4
Warsaw 108
Whittington, Aven Jr 56–7
Wittenberg, East Germany, Agrochemical Centre 80
World Trade Association, GATT talks 64
Wroclaw (Breslau), Poland 98–9

Zarrantin LPG, East German border 78–80
Zieliniski, Jan 104–5